A Journey from Dust to Consciousness

Vincent J. Hyde

BALBOA.
PRESS
A DIVISION OF HAY HOUSE

Balboa Press books may be ordered through booksellers or by contacting:

Balboa Press
A Division of Hay House
1663 Liberty Drive
Bloomington, IN 47403
www.balboapress.com.au
1 (877) 407-4847

Print information available on the last page.

ISBN: 978-1-5043-1247-9 (sc)
ISBN: 978-1-5043-1248-6 (e)

Balboa Press rev. date: 03/29/2018

Acknowledgements

The author acknowledges that many of the scientific facts mentioned in this book may be found in encyclopaedias and science textbooks. It is possible that due to the quantity of books and encyclopaedias available, the wording used in the book is similar to that used in other encyclopaedias and textbooks. It is not the intention of the author to infringe upon any copyrights. The author has stated the scientific facts to suit the script and story in the book and is thankful that the encyclopaedias and science text books exist to strengthen his point of view.

This book has been edited by Balboa Press. As a result, there are some minor differences to the earlier book 'Heaven and Earth' by the same author. Example the number of elements have changed from 103 in Heaven and Earth to 118 in this book. This is because 15 new elements have been recently discovered.

In this book the word 'lifeform' has been made one word to describe animate organisms that contain one or more living cells. Some text books may refer to this with two words 'life form'.

Explanation for Upper Case Used in the Book

Upper case has been used to make it easy for the reader to follow the text for the following reasons:

1. A single word might have a vague meaning if considered by itself. However, in the context of the book, it has a particular meaning; hence it is shown as upper-case text. For example, the word "Field" can have a meaning associated with either the Fields of Knowledge or a Field associated with a star system, depending on the context.
2. A two-word term might be required to be considered together and not separately. For example, the two-word term "Black Hole" is associated with objects that exist in the centres of galaxies and produce high energy radiation.

No image information in this book

Image information on such objects as galaxies, stars, planets, moons, and machines has not been provided in the book. Readers wishing to obtain image information may google the information directly as required. This will allow the readers to access the exact Field and piece of knowledge they require.

Author's Comment on the Word "Star"

In this book, the word "star" has been used in two contexts that have been fused together as part of the story.

In the early chapters of the book, the word "star" describes the dust that orbits a Black Hole to form the objects that fuse hydrogen in their cores to form the elements and generate the Field and the energy for their planets and moons.

In later chapters, the word "star" describes successful human consciousness in the Fields of Knowledge.

The reader should read the book to determine the meaning of the word "star."

A Brief Description and Book Intent for Browsers

Most people live their lives working in the Fields of Knowledge to earn money to buy essentials like food, water, transport vehicles, services, and shelter. This novel describes the bigger picture and includes the routine of daily life.

The book describes how the dust of the universe is transformed into billions of unique galaxies with unique stars, planets, and moons.

The stars go through cycles of birth, middle age, old age, and death and spend their relatively long lives making the elements and providing the energy for their planets. The planets and moons are places where temperatures and pressures allow for the development of varying conditions and climates to allow for many chemical reactions to occur.

Life exists on Earth, where the sun provides a constant supply of energy. Lifeforms are made up of cells, which contain chromosomes and genes that contain instructions for automatically building complex organs and carrying out processes necessary to allow lifeforms to be conscious for a relatively short period of time.

Reproduction, whereby two independent cells can combine to form a new and unique offspring, could be seen as the work of a Creator, though not everyone will agree with this statement. Reproduction is certainly a nonhuman way of producing goods. Human processes produce goods at a high production cost.

Human consciousness is unique when considering the other lifeforms on Earth. Through the development of language, communication, and data storage, human beings have created millions of Fields of Knowledge.

The Fields of Knowledge are stored in libraries and on the internet, where they are accessible to all human beings capable of reading.

Human stars work and are capable of developing the Fields of Knowledge and even creating new Fields of Knowledge for future human beings. Even when they pass away, human beings may be remembered through the Fields of Knowledge and may influence the development of new stars.

The Fields of Knowledge seem to be driving human civilization to utilize large amounts of energy and to create machines to do work. The dangers of using fossil fuels and non-renewable resources is considered, and the use of a worldwide power network using solar and wind power is offered as a possible solution.

Machines are seen as significant in taking consciousness to other galaxies. Using machines is seen as a relatively save way of exploring and transforming the dust on other worlds.

This novel is intended to be an open system. Those that choose to view the concepts as science may do so. Those that choose to view the concepts as science fantasy may do so. Those that choose to view the concepts as religious may do so. Though the book favours a Creator, it is up to the reader to choose whether to pray or not to pray.

The message of the book is that, whatever you choose, it is wise to use the cells given to you to work hard in your chosen Field so that you can become a star. Then, like the stars, you can shape the Field and perhaps create new Fields for future stars.

Contents

Introduction

Every year, forty days before Easter Sunday, Christians celebrate Ash Wednesday. During the hourly celebration, a priest blesses the ashes and then places them on the foreheads of individuals, saying the following words: "Remember man that you are dust, and into dust you shall return." These words have been handed down for generations and are believed to be from the Creator.

These are remarkable words, and it is remarkable that of all the species on Earth formed from the dust—including the trees, fish, reptiles, amphibians, birds, insects, and mammals—only human beings have the consciousness to understand the meaning of the words. On Earth, only human beings have written language and are capable of storing data in paper print and electronic forms to create the Fields of Knowledge. Our human senses, together with the accumulated knowledge of the generations, form a picture of the world. It is a picture that can change over time with more knowledge. Human beings can add to the Fields of Knowledge, changing them for future generations and perhaps forming new Fields of Knowledge. In a similar manner, the stars can change the galaxies, creating new fields for new generations of stars. It is an amazing process that leads to a very interesting universe filled with fields that spawn ever-changing stars and minds.

Just as the stars were formed from the dust clouds, nebulae, within the galaxies, human beings are formed from the dust on Earth. Both processes are a remarkable transformation that makes the dust more energetic and alive. Eventually the dust runs out of energy and returns to its normal state of being dust. It is as if the dust gets energy and becomes ordered; then, as energy is lost, disorder returns.

In this book, we will see:

a) How the dust is transformed into galaxies with stars
b) How the dust is transformed into life and consciousness on Earth.

The universe, with its galaxies and stars, contains enormous power and energy, operates on time scales that span billions of years, and encompasses enormous forces and Fields that are easily understood by human intelligence through the Fields of Knowledge.

Like the galaxies, the Fields of Knowledge are diverse, including all forms of human work and activity.

The atoms that come to life, in a sense, on worlds around the stars contain enormous power and energy. They can transform themselves from solids to liquids and to gases, and vice versa. The atoms can automatically use energy from the star to form molecules and chains of molecules. The atoms can convert energy into life if the conditions around the star are suitable or if the Creator allows it to come about.

The genetic code in all lifeforms allows for reproduction and the complex process of allowing for millions of chemical reactions that occur automatically within the bodies of all plants and animals. This justifies the religious belief in a Creator. Metabolism within the bodies of plants and animals converts food into energy without the individual lifeform knowing it is happening.

Life can become conscious of its surroundings via sensors, in the case of plants, or via a brain and sensors, in the case of animals. Earth's surface has varying temperatures and pressures that result in different climatic conditions. This gives rise to different biomes, where plants and animals live in mutually beneficial relationships. Human consciousness is unique and capable of transforming a planet like Earth by utilizing the available energy.

Data Storage in books and electronic forms, together with an excellent education system, has made the Fields of Knowledge very reliable, accurate and easily available in libraries and on the internet.

Human consciousness is constantly developing via the many Fields of Knowledge all around us. Today electrical energy is used to power our houses and operate the many machines and appliances we possess around our homes. Our cities are illuminated by lights, which enables safe night-time access around the planet. Food and water are readily available in cities and towns all over the world. Mechanical vehicles move on our streets, transporting people and goods across the cities and countries. Trains, trams, ships, and planes transport people and goods around Earth at all times. Machines help us to grow food and package goods for transport. Machines help us to build and maintain roads, tunnels, and even the buildings we live and work in. All these machines use energy and are controlled by signals and computers under human supervision and programming. The development of automation has given machines some degree of freedom, and future developments certainly look promising with regard to a brighter future for everyone.

Through the Fields of Knowledge, our machines are capable of exploring other worlds and sending data and pictures back to Earth. Surely this is an indication that human consciousness is capable of exploring the dust on other worlds. If suitable, human consciousness may create machine life on other worlds using the energy available from the stars of other worlds. Machines can easily adapt to extreme conditions likely to occur on other worlds in other star systems. In this way, human consciousness can exist on other worlds in other galaxies. Using our machines and the energy from the stars, we can take human consciousness to all the galaxies in space.

The rapid development in the Fields of Knowledge over the last three centuries has resulted in some degree of pollution, as our machines are using increasing amounts of energy, though they themselves are energy efficient. As the machines are a product of the Fields of Knowledge developed by human consciousness, it appears to be of significant importance to use natural resources wisely, so that, they do not damage the environment and the biomes that exist around us. Using natural resources that are renewable and non-polluting is important, because it appears that our machines will be with us for a long time and perhaps forever.

Regarding the use of energy on a large scale, there are always two views, an optimistic one and a pessimistic one, and perhaps only the Creator knows what the final result is going to be.

1

DUST

THERE WERE CHURNING oceans of dust everywhere in space. The simplest of atoms, the hydrogen atoms, began to form everywhere. Hydrogen atoms are the simplest configuration possible, consisting of a proton surrounded by a single electron. This atom could form with minimum energy, as the negative charge balanced the positive charge and created an equilibrium. The first element in the periodic table, with the atomic number 1, was everywhere. To make the other elements with higher atomic numbers, the hydrogen nuclei had to be forced together. The rule that positive nuclei repel each other made it impossible to create the heavier elements with higher atomic numbers. To make the elements with higher atomic numbers, there needed to be rules and laws whereby the hydrogen nuclei could be squeezed together to form stable heavier elements. This process could only be carried out by nuclear fusion in the middle of a star.

In the oceans of dust, there was complete chaos and total randomness. Time, as we know it, did not exist.

Fields, rules, and laws came into existence to build the heavier elements in the centres of large stars, forcing them to expand and then collapse into extremely dense and rapidly rotating black holes. The black holes became the seeds for the formation of the galaxies.

Some of the fields created are the following:

- the gravitational field
- the electrical field

- the chemical field
- the temperature field
- the pressure field
- the magnetic field

All the fields have rules and laws that impose order on the matter within the field. The hydrogen atoms, with their single protons and single electrons, are squeezed in the centres of the vast dust clouds by gravity to create helium, the element with atomic number 2, resulting in a release of energy. Each sea of atoms is subjected to enormous pressure as the gravity field pulls matter in and the created energy pushes matter out. Eventually an equilibrium is reached and the large stars become stable.

However, to balance the inward pull of the gravity field, energy has to be expended in the conversion of hydrogen into helium. Then, as the hydrogen fuel is consumed, the stars have to fuse helium atoms to make carbon. When the helium fuel is consumed, the stars fuse carbon atoms to form oxygen. When the carbon fuel is consumed, the stars fuse oxygen atoms to form iron. Iron cannot be fused, and the stars run out of fuel when iron forms in their cores. While iron is forming, the stars heat up and expand, and after iron forms, they collapse. This collapse of a star results in a very small, dense, and rapidly rotating black hole that becomes the seed for the formation of a galaxy. The newly formed atoms are thrown into space, where they are quickly swept into orbit around the dense and rapidly rotating black hole. The sea of hydrogen atoms is transformed into a galaxy composed of clouds of gas and dust rotating around a black hole. From a sea of dust moving randomly with no sense of time, the rules and laws created dust and gas with heavy elements rotating rapidly around a very dense bit of matter. This rotation gives the sensation of time because stars live and die as they orbit the black hole within their galaxy.

The stars go through the process of birth, middle age, old age, and death in a process that takes billions of years. Throughout the galaxy, the dust has been converted not only into stars but also into tiny atoms that are distributed in space by the movement and death of the stars. Also, in living and dying within a galaxy, the star transforms the galaxy, creating exotic spiral arms and changing the galaxy's shape and form. This process makes each galaxy unique with its own identity.

The fields form new stars within the galaxy. The new stars have planets with an abundance of elements, and some of the planets have moons with an abundance of elements. Thus, within a typical galaxy, the fields produce smaller stars that fuse elements at a slower rate and live for billions of years. The energy from the stars falls on the planets and moons rotating about the stars, creating very varied environments with diverse temperatures and pressures.

The variability that has been built into the design is truly amazing. It appears simple but is extremely complex in the way rules and laws impose an underlying order and overall equilibrium. The initial dust clouds are very similar, with hydrogen forming easily in a sea of matter. Then, with the rules and laws, the dust is transformed into an ordered system of matter rotating about a black hole in a galaxy filled with energy, where stars go through a cycle of birth, middle age, old age, and death. The stars transform the galaxy, giving it a unique identity. The stars provide fields for their families of planets and moons, making each star unique and each planet and moon also unique.

With so many environments, the design provides the scope for the dust transforming into life.

Earth is a planet rotating and revolving around the sun in the unique solar system of the Milky Way Galaxy. We humans have been provided with brains, sensors, air intake and exhaust systems, blood-circulating systems, energy intake systems, and waste elimination systems, along with skills like building things, storing data, and communicating. If we are dust, as the Creator says, then is it amazing how the dust can become so complicated as to become us. In human beings, the dust has been transformed from a sea of hydrogen atoms into beings with consciousness that can see and understand the universe.

2

GALAXIES

LOOKING UP AT the night sky from Earth, we see the stars moving across the sky from east to west. However, telescopic observation informs us that what we are actually seeing is the Milky Way Galaxy with one hundred (100) billion stars spinning around a black hole at the centre. The central bulge of the Milky Way Galaxy is twenty thousand light years across and three thousand light years thick. The whole galaxy is one hundred thousand light years across and one thousand light years thick. The Milky Way Galaxy is a typical spiral galaxy, with the solar system containing the Earth located in one of the arms of the galaxy. It takes approximately two hundred and twenty (220) million years for the sun to make one circuit around the central Black Hole of the Milky Way Galaxy.

The galaxies appear to be regions where the fields, rules, and laws can act on the dust to transform the dust into stars, planets, and moons. The transformation of dust into a star results in the creation of a new field with rules and laws. The fusion of atoms in the core of the star results in a release of energy, which creates all manner of landscapes and climates on the planets and moons in the star system. This allows for all manner of chemical reactions and compounds to form. As a star like the sun may live for nine billion Earth years in a very stable state, it is possible for the chemicals produced to follow rules and laws that encourage the growth and development of highly organized life. If this is true, then galaxies are places where dust may be transformed into life over long periods of time.

The number of galaxies in the universe is estimated at five hundred (500) billion. Galaxies are known to interact with each other, pulling each other, changing shape, and sometimes even combining to produce new fields and stars. The galaxies do not appear to be intelligent; they merely appear to be following the rules and laws that are configured into the fields to transform the dust into all varieties of stars, planets, and moons.

Galaxies are places where stars live their long and perhaps interesting lives in fields, according to rules and laws. The stars are born within the galaxies in thick clouds of gas and dust known as nebulae. "*Nebular*" is the Latin word for "cloud."

A typical galaxy has one hundred (100) billion stars in various stages of their lives. Some stars are young, bright, and blue. Other stars are middle-aged and yellow, like the sun, while others have reached old age and emit less energetic red light.

The stars are extremely significant for the galaxy. The stars are objects that are alive and that produce energy that lifeforms in other galaxies can recognize with their senses. From early times, human beings realized that there were other objects outside of Earth because their cells had formed eyes capable of detecting the visible light emitted by the stars in the galaxies.

The fusion reactions occurring in the cores of stars form a variety of elements through the systematic conversion of hydrogen into heavier elements. The process fuses hydrogen into helium, helium into carbon, carbon into oxygen, and oxygen into iron, giving out large quantities of energy during all stages. The rules and laws do not allow this process to continue forever. Hence, the stars go through the process of birth, middle age, and death within the galaxy. In death, the elements formed within the star are released into the galaxy to enrich the dust clouds for the creation of future new stars, planets, and moons.

The energy from the star must fall on the planets surrounding the star, and this affects the surface features and atmospheres of the planets. Planets, in turn, can influence the surface features and atmospheres of their moons.

Objects with mass exert force on other objects with mass. This means that massive objects can crash into each other, forming objects with even greater mass. However, not all objects crash into each other. Some tend to revolve around each other, again based on rules and laws, resulting in a reasonably stable basic unit known as a solar system. Planets and moons that form around the central core have various characteristics and compositions based on their location in the star system and the elements available in the initial cloud. Obviously, as the process of energy turning to matter and then matter turning into energy continues, the dust clouds begin to contain all variety of elements, making the galaxy richer and granting it the ability to create more exotic planets and moons.

A star exists for billions of years because nuclear fusion allows the star to burn its fuel at a relatively slow rate and prevents gravity from pulling the star inward. Throughout middle age, the star maintains a constant size as equilibrium exists between gravity pulling it inward and nuclear fusion pushing it outward. In old age, the star becomes unstable, as nuclear fusion cannot be sustained. The star expands outward until gravity pulls it inward, making it a very dense piece of Dark Matter.

The fact that the star can remain stable for billions of years is very significant, because it allows for a variety of possibilities for the planets and moons in the star system. As we know, water-based life exists on Earth in a so-called Goldilocks Zone about the sun, where the temperature is just right—neither extremely hot nor extremely cold. However, what is interesting is that if the star produces less energy, the Goldilocks Zone may shift inward. Also, if the Star produces more energy, the Goldilocks Zone may shift outward. Nothing appears to stay the same forever—not even the Goldilocks Zone about a star. The ability to change its Goldilocks Zone allows other planets and moons about the star to have favourable temperatures. This means that areas around a star that are not favourable for life may become favourable at a different point in time, allowing new resources about the star to become capable of supporting life. However, as observed with our solar system, changes in the Goldilocks Zone about a star occur very slowly, making it very difficult for lifeforms living for one hundred years to detect or panic about them.

Galaxies contain billions of stars. Each star produces an abundance of elements in its core and then distributes these elements in Space for other future stars. When combined with the relatively long lifetime of the star and the fact that the energy distribution allows for the creation of Goldilocks Zones, this allows a variety of possibilities, including life, for the planets and moons that revolve about the star. As each galaxy contains billions of stars, there are many possibilities for interesting chemistry within each galaxy.

The nature of the elements formed in the star, and the ability of the elements to form all manner of complex molecules and compounds on planets and moons, is again controlled by very definite rules and laws.

The massive stars that exhaust their fuel and expand before collapsing into relatively dense dark matter have extremely high rotation and extreme magnetic energy, which causes matter to swirl around them. All Stars and dark matter emit energy in the form of electromagnetic waves at varying frequencies. The wavelengths, or the distance between any two peaks of the wave, fall into the following ranges:

- radio waves: from 10^4 to 10^{-2} metres
- microwaves: from 10^{-2} to 10^{-4} metres
- Infrared waves: from 10^{-4} to 10^{-6} metres
- visible waves (red, orange, yellow, green, blue, indigo, and violet): from 10^{-6} to 10^{-7} metres
- ultraviolet waves from 10^{-7} to 10^{-8} metres
- X-rays from 10^{-8} to 10^{-14} metres
- cosmic ray waves from 10^{-14} to 10^{-16} metres

These waves give an indication of the energy released, and spectroscopic studies can indicate the elements present in the various stars of a galaxy. Both stars and dark matter release energy and radiation in accordance with very definite rules and laws.

All the above complexity of dust being transformed into matter and energy makes each galaxy unique, with a characteristic appearance, size, and shape. Astronomers have identified four types of galaxies—namely spiral, elliptical, irregular, and barred.

Spiral Galaxies are easy to identify by their sweeping arms, which contain gas and dust and make new stars. An important subclass of the Spiral Galaxies are the Barred Spirals. The Barred Spirals have roughly oblong-shaped centres and may have undergone collisions with other small galaxies.

Elliptical Galaxies comprise mostly older stars contain little gas to make new stars. A subclass of the Elliptical Galaxies are the Dwarf Elliptical

Galaxies. These galaxies are numerous; however, they are difficult to see, as they contain dim stars. Astronomers think that these ball- or oval-shaped galaxies may have formed early in the history of the universe.

Irregular Galaxies are small and shapeless. Many Irregular Galaxies are still actively making stars. The small Magellanic Cloud is an Irregular Galaxy that is being distorted by the Large Magellanic Cloud and the Milky Way Galaxy.

3

STARS

MOST STARS ARE born in vast dust clouds located in the galaxies. Generally, the same dust cloud that forms the star gives rise to planets, moons, asteroids, comets, meteorites, and others debris all revolving in the new Field created by the new star. The star and its family of matter form a Solar System that is very dynamic, with matter colliding and increasing the mass of the relatively stable planets and moons. In time, meteor bombardment and collisions grow less frequent, until after long periods of time (billions of earth years) one sees a relatively stable star with stable planets and moons, with occasional meteor impacts roughly every one hundred Earth years. The Star System is like a basic unit for the galaxy. Typically, one hundred (100) billion star systems make up a typical galaxy. The star revolves around the centre of galaxy with its family of planets, moons, and associated matter. Stars go through cycles of birth, middle age, old age and death. In old age and death, stars produce the heavy elements that are distributed into the galaxy and are essential for the creation of new stars with planets and moons, where the elements may be able to form conscious lifeforms.

Most of the mass of the initial dust cloud, perhaps 99.9% of it, forms the star. The remaining 0.1% of the mass forms the planets, moons, asteroids, comets, and other debris. The star is significant for producing energy, the elements and the Field. However, the planets and moons are significant for forming the climates that can support lifeforms and consciousness.

Stars within a galaxy form in a similar manner to the Black Hole that formed in the centre of the galaxy. However the new stars are far less massive and have smaller Fields to support their family of planets, moons, comets, meteorites, and debris revolving around them.

The element hydrogen is abundant within each galaxy. Hydrogen forms readily, requiring minimum energy. Each galaxy has plenty of hydrogen rotating in vast dust clouds about its central Black Hole. The element hydrogen is the simplest element in the Periodic Table of elements. Hydrogen consists of a single proton with positive charge and an electron with negative charge orbiting around the proton. The cloud of hydrogen becomes spherical as the Fields act on the hydrogen over time. Eventually the hydrogen-rich central region of the spherical cloud is squeezed, and two protons are pushed together to form the element helium and release energy. This process of nuclear fusion in the core of a star is the result of a couple of fields with well-defined rules and laws:

- an electrical field of protons, electrons, and neutrons
- a gravitational field pulling the Star inward
- a pressure field caused by matter being pulled inward on a very large scale
- a temperature field
- an energy field pushing outward as atoms are fused together, with the loss of mass resulting in the creation of energy

To prevent the new star from collapsing, more and more matter has to be fused in the core of the star to create the outward push of heat and energy. Neutrons with neutral charge are formed when electrons are squeezed into protons.

In the young star, the fields created fuse hydrogen to helium. When the supply of hydrogen diminishes, energy must be created by fusing helium to form an element with a higher atomic number—namely carbon. When the helium is diminished, energy must be created by fusing carbon to form an element with a higher atomic number—namely oxygen. When the supply of carbon is diminished, energy must be created by fusing oxygen to form an element with higher atomic number—namely iron. The process of fusing elements is extremely long, which allows the star to shine for billions of years. However, when iron is created in the cores of massive stars, there is an energy problem, because the rules of fusion do not allow iron to fuse to form an element with higher atomic number. The result is that the star heats up and expands when it fuses elements higher than helium. However, when iron is created in the core, there is no further energy produced, because the element iron cannot be fused in the star's core. The star is pulled inward by the gravitational field, causing a massive explosion in which most of the one hundred elements created are thrown out into the galaxy. After this, a very dense core spins rapidly, transforming the Space within the galaxy by creating exotic arms that give a unique shape to the galaxy.

Time starts with the massive explosion and the creation of the Black Hole in the centre of the galaxy. The enormous and powerful Field created by the Black Hole moves matter around it. In time, the rules and laws allow all manner of smaller stars to form in the galaxy. Stars begin life in clouds of gas and dust called nebulae. Inside nebulae, the gravitational fields create dark clumps of matter as a result of differing rotational speeds around the central Black Holes. If a dark clump of matter is squeezed until its temperature reaches ten million degrees centigrade, the hydrogen in the dark clump fuses to form helium, and the energy released causes the new star to glow.

Astronomers work out the size of a star from its brightness and its temperature. The brightness of a star depends on its mass. The mass of a star is generally controlled by the size of the dark clump of matter that is brought together in the new gravitational field about the Black Hole. Stars with the mass of the sun are medium-sized stars. The dark clumps of matter can range from one hundred times the mass of the sun to 6 per cent of the mass of the sun.

Stars fall into three basic categories:

1) Large white stars bigger than the sun
2) Medium-sized stars the size of the sun
3) Small red dwarfs or brown dwarfs the size of the planet Jupiter

Large stars are hot and white. Large stars make energy fast and have high temperatures. They produce fusion reactions that result in heavy elements being formed in their cores. Super large stars with masses five hundred times that of the sun can produce heavy elements involving the fusing of carbon to form iron. These stars end up using their fuel and collapsing in a supernova explosion. This results in the formation a Neutron Star typically twenty kilometres across, with the mass of the sun formed from the central core of the star. The Neutron Star is super dense, with a crust of iron and similar elements. Neutron Stars spin very rapidly, beaming out regular pulses of radio waves. They were first discovered in 1960 and are also called pulsars because they produce radio waves at regular short intervals—typically once every second. X-rays are created as material is squeezed by the very strong gravity and powerful magnetic field of a neutron star.

Medium-sized stars like the sun undergo fusion reactions at a much slower rate than the large stars. The fusion of hydrogen into helium pushes energy out while gravity pulls the star inward. The star might shine in a steady equilibrium state for seven billion years. When these stars run out of fuel, they expand, becoming what astronomers call Red Giants. If the sun were to become a Red Giant, its outer layers would probably reach the orbit of Mars. The Red Giant blows off its outer layers and collapses to become an object the size of Neptune with a mass equivalent to that of the sun, known by astronomers as a while dwarf.

Variable Stars are stars that vary in brightness. The variation in brightness is caused by the star flaring. Pulsating Variables are stars that expand and contract. Cepheid Variables are big, bright stars that pulsate with energy, flaring up regularly every one to fifty days. RR Lyrae Variables are yellow super giant stars that flicker and vary in brightness as their fuel runs down.

Binary Stars are double stars that are held together by one another's gravitational fields.

Small stars have a relatively slow rate of fusion and are relatively cooler than the other two categories of stars. They are the size of Jupiter and glow with approximately 5 per cent of the sun's brightness. These stars live for approximately two hundred billion years. They might end their lives by forming a crust and glowing very faintly, becoming what astronomers call a brown dwarf. Finally, these stars might completely cool off as the whole star solidifies, becoming what astronomers call a black dwarf.

The gravitational field that collapses the large gas cloud to form the rapidly rotating star also produces a solar system of rotating planets, moons, asteroids, and comets. This makes each solar system, planet, moon, asteroid, and comet unique.

The star provides a new gravitational field for the planets and associated objects, depending on its mass. It also provides an energy source for the planets, the strength of which depends on its mass and how it fuses its fuel. The energy source provides the climate for the planets.

The planet provides a new gravitational field for its moons and associated objects, as well as an energy source for its moons; the strength of both of these depends on the planet's mass. Generally, the energy is tidal energy in the form of heat. Additionally, the planet exerts rotational influence on its moons.

The original field that collapsed the large gas cloud provides distribution of elements to the planets and moons, determines the positioning of the planets and moons relative the star, provides the asteroids and meteorites that impact the planets and moons, and is responsible for the rotational energy of the planets and moons.

Large planets like Jupiter and Saturn have their own gravitational fields that control their own moons, allowing for a variety of conditions that are not directly dependent on the star.

The star converts the dust into an energy source and makes the elements by fusing hydrogen in its core. A star is too hot to contain solids and liquids on its surface. Matter on the surface of a star can exist only as a plasma. The planets and moons revolving around the star can have a range of temperatures and pressures, which allows the elements to form a variety of compounds.

The fact that medium-sized stars are designed to live in a stable state for about nine billion years, while small stars are designed to live in a stable state for approximately two hundred billion years, implies that the Creator has provided a relatively stable source of energy to act on the elements distributed on planets and moons to transform the elements into a variety of compounds over long periods of time. Obviously, given the right conditions, the dust may be transformed into life, and finally into life with consciousness. The energy of the sun results in various climates on the planets; a typical example is Venus, where surface temperatures are higher than the boiling point of lead. On Mars the surface temperatures are so cold that water is frozen in the desert-like soil.

We are very fortunate to have a planet like Earth, where the elements formed plenty of liquid water to encourage chemical reactions. Also, water can retain the energy of the sun and can exist as a solid, liquid, and gas. In addition, the tides and waves would have assisted in mixing the elements and forming numerous compounds to encourage the development of life from the dust.

Though the star's energy and the elements are extremely important ingredients for life, a dissolving agent like water is necessary to mix the elements and create life.

However, for consciousness, one needs life to have a high degree of intelligence with some sort of brain and sensors to gather information and to interact with one's surroundings, while also possessing the ability to reproduce. In short, one needs more than a star's energy; one needs life to possess some sort of genetic code to store data and automatically transfer it to the next generation. The genetic code can be seen as a system of rules that can be changed as conditions surrounding the lifeform change. In this way, lifeforms may adapt to changes taking place in their surroundings. Environmental change appears to occur often, and it is essential to have a system that is flexible enough to allow for such changes, which generally occur slowly. The genetic code is unique for each species of plant, tree, fish, amphibian, insect, and animal. The genetic code is complex, controlling the actions of cells within the lifeform. The code seems to be programmed and is not directly under the control of any lifeform. It appears to be the work of the Creator, though not everyone will agree with this.

It is, however, interesting to note that conscious life has to take in sunlight, food, air, and liquid available on a planet's surface and has to have a body and organs capable of transforming these items into energy to function and be alive. On Earth, this task is done by lifeforms having cells with chromosomes and genes. The genes within each cell have instructions to break down complex molecules eaten by the lifeform, and to form simple molecules and energy. The lifeform then has an energy transportation network and a waste disposal network to distribute energy to all cells and eliminate waste back onto the planet's surface. This system allows a lifeform on Earth to exist for about one hundred Earth years, which is a short time compared to a star's lifetime of a few billion years. In a dynamic star system, a relatively short lifetime for a lifeform may be a blessing, because, the lifeform may witness relatively few disasters on the planetary scale and star scale. On Earth, disasters on a planetary scale include hurricanes, floods, volcanic eruptions, meteor strikes, and earthquakes.

Finally, for human-type consciousness to emerge in a star system, one needs the lifeform to be able to utilize the energy of the star. The lifeform has to stop searching for food on its planet to make energy for its own existence. The lifeform has to find other sources of fuel that can be transformed into electrical, mechanical, nuclear, and chemical energy on a planetary scale. Also, the lifeform has to find the resources on a planet's surface to develop machines that can use the energy available from the star to do work on a planetary scale.

4 ENERGY AND THE ELEMENTS

THE GALAXIES ARE places where the stars are born and live their lives. The stars are doing three main tasks:

1) Producing energy
2) Making the elements in their cores
3) Creating fields for the planets

In the earlier chapters, I explained how the stars produce energy in their cores by a process known as nuclear fusion. This energy produced by the star must travel out through Space to the planets.

The planets and moons contain the natural stable elements in their soils. The types and amounts of these elements present are partially dependent on the initial gas cloud from which they formed and partially dependent on how much they have cooled over time. Planets and moons have their own internal heating, which occurs in their cores. This internal heating is not strong enough to squeeze protons together to form the heavy elements, as the stars do. In some cases, the energy of the star can create useful reactions among the elements to produce, and certainly support, life in accordance with rules and laws. The rules and laws seem to drive life in the direction of consciousness, until eventually the dust becomes conscious and is able to transform and even control the energy of the star and the elements on its planet or moon. This is when the Creator reminds the conscious beings that "they are dust and into dust they will return."

To be conscious, as human beings understand consciousness, one must possess a brain connected to sensors, such as ears, eyes, a tongue, a nose, and skin as a minimum. Then, an individual identity can understand or be conscious of the surroundings. The sense of vision enabled human beings to realize that the stars, and in particular the sun, are producing enormous quantities of energy in a steady manner. The sunshine on Earth, together with water, makes plants grow. Energy from the star is being converted into growing grass, plants, and trees, which are a source of food for hungry lifeforms.

All the stars, planets, and moons have an escape velocity. To escape the gravitational pull of the star, an object must reach the star's escape velocity. In the case of the sun, its escape velocity is approximately three hundred thousand kilometres per second. As the sun or any other star fuses hydrogen in its core, a stream of high-energy photons is released. This energy raises the temperature of the star, and the surface temperature becomes over 5,000° C. This enables the photons to reach the speed required to escape in all directions into space. As the photons leave the star's surface, they have various energies. The higher the energy, the greater the frequency and the shorter the wavelength. Thus the most energetic gamma rays leave at very high frequencies and very short wavelengths; they are followed by X-rays at high frequencies and short wavelengths. Next is ultraviolet light, which is followed by visible light, microwaves, and radio waves, in that order.

Black Holes have extremely high densities, and visible light cannot escape from their surfaces. This is why it took scientists so long to discover their theoretical existence. Quasi-stellar radio sources, or quasars were discovered recently. These are the most intense sources of light in the universe. Quasars the size of a solar system glow with the brightness of one hundred galaxies. These objects are very distant and seem to always be at the heart of a galaxy. It was theoretically believed that the energy produced by quasars was produced by a Black Hole with a theoretical mass of one hundred million suns.

Thankfully, normal visible stars do not display such extreme behaviour. The hot white stars have many photons leaving the star's surface at the higher blue frequency, and we see these as blue or white stars. The medium-sized stars, like the sun, have photons leaving at the yellow frequency, and we see these as yellow stars. When a yellow star runs out of fuel, it expands, becoming a Red Giant. Astronomers have observed that this is what is happening to the Star Betelgeuse. Its increase in size makes its outer layers cooler, and the red frequency becomes the dominant escape frequency for the photons. The small stars, like Arcturus and Antares, glow red because the less energetic photons vibrating at the red frequency easily escape the small stars' weak gravity fields.

To get an idea of the energy produced by a typical galaxy like the Milky Way, we must recall the energy produced by the sun and then realize that the Milky Way contains an estimated one hundred billion stars. If this is large, then the energy produced by the universe is enormous, because the universe is believed to contain an estimated 500 billion galaxies.

We know that stars have planets rotating and revolving in the fields around them. Also, many planets have moons rotating and revolving around them. We know that the planets and moons contain the elements created in the cores of stars and distributed into Space in supernovae and other star emissions. This process is the Creator's way of converting the dust into energy and elements. Obviously, an essential feature of such a design must be for the energy to act on the elements to transform them into some form of consciousness or to make the dust animate. The alternative is a system that has elements remaining in inanimate form.

All matter in the universe is made up of various combinations of the 118 known elements. The elements are substances that cannot be broken down into simpler substances by chemical methods. There are 94 naturally occurring elements and 24 synthetic elements.

The elements are classified as metals and non-metals.

Metals are shiny solids that conduct electricity. Most metals melt at high temperatures. Metals are malleable, which means they can be hammered into various shapes and sizes. Most metals are also ductile, which means they can be stretched without breaking. Most non-metals melt at lower temperatures than metals, and many are gaseous at room temperature.

Generally, elements can combine with other elements to form compounds. Chemical reactions can also be used to break down compounds and free the elements they contain.

The synthetic elements are formed on Earth in nuclear reactions using natural elements. The synthetic elements are so unstable that they decay and fall apart quickly, often in minutes or even fractions of seconds.

Uranium is the heaviest naturally occurring element. It has 92 protons, 92 electrons and 146 neutrons. Uranium is formed during some sort of stellar supernova explosions. Elements heavier than uranium are unstable. Their nuclei burst apart because the forces that draw protons together are not strong enough to overcome the repulsion between their positive charges.

The 118 known elements form a pattern when arranged in increasing atomic number. Fortunately, human beings have worked out the pattern, and it makes a very interesting table called the periodic table. On this table, elements are arranged in order of increasing atomic number, with similar elements grouped together. The periodic table lists the 118 elements formed in the cores of stars and distributed to the planets and moons in eighteen vertical columns, or groups, and seven horizontal rows, or periods. The elements are arranged so that their atomic numbers increase from left to right through the horizontal periods. An element's atomic number is the number of protons in its nucleus and the number of electrons orbiting the nucleus. Elements in the same group have similar properties. The chemical properties of an element depend largely on the number of electrons in the outermost shell of the element. The electrons are arranged in shells around the nucleus.

On the planets and moons, given the right conditions, the rules and laws allow element interaction. Such interaction depends on the electrons in the outermost shells only. This allows for the formation of many very interesting compounds that can exist in stable form on a planet's surface.

As an example, consider the water molecule, hydrogen oxide (H_2O), which consists of two atoms of the element hydrogen combined with one atom of the element oxygen. Oxygen has the atomic number 8; as it has two electrons in its first shell and six electrons in its second shell. That leaves space for two electrons to fill its second shell, which can hold eight electrons. Hydrogen has the atomic number 1, as there is only one electron in its outer shell. That leaves space for one electron to fill its outer shell, which can hold two electrons. Hence, two hydrogen atoms can combine with one oxygen atom to form a water molecule. The rules allow this combination to occur. Obviously, if conditions are satisfactory, and if there is plenty of oxygen and hydrogen available on a planet's surface, water will form in great abundance. In addition, water is a polar molecule because the oxygen has a slight negative charge while the hydrogen has a slight positive charge. These charges attract water molecules to each other, which makes water an excellent solvent. As an example, everyone knows how sweet a cup of tea tastes with two teaspoons of sugar dissolved in it. It makes some people have a second cup of tea.

The other factor is the state of matter, which is dependent on a planet's energy from the star and its internal energy from its own internal core. Matter on a planet's or moon's surface can exist in three basic forms: solid, liquid, or gas. On Earth, for example, the average temperature is 25° C. At this temperature, water exists as a liquid. Hence, water gathers in low-lying seas and oceans. At the poles the temperature drops to 0° C and heat energy is removed from the water, converting the liquid water into solid water, or ice. In the equatorial regions, the sun heats the liquid water, and the added energy causes large quantities of the surface water to evaporate, forming clouds, which drift towards the land. If they strike a mountain or raised land, the clouds of water vapour rise. The resulting cooler condition removes heat energy from the water vapour gas, converting it to a liquid, and it falls as rain. The rain dissolves minerals and salts out of the rocks on the mountain and carries them via rivers to the seas and oceans. Even solid ice is known to flow over the land as glaciers, carrying rocks and minerals around with it. This ability of water to move around the land, powered by energy from the sun, is crucial in bringing all sorts of minerals and elements into the oceans. Approximately 70 per cent of Earth's surface is covered with water. The energy from the sun keeps the liquid constantly moving and full of energy, which is stored as heat within the water. Water also exerts pressure on the thin ocean floor, getting elements and chemicals directly from Earth's internal core. All these factors make it ideal for simple lifeforms to emerge from the energy in the water, following the rules that allow the elements to bond with each other and form compounds.

In any substance, particles are in constant motion. This energy, called kinetic energy, increases with temperature. Whether a substance is a solid, liquid, or gas depends on the balance between kinetic energy and the forces of attraction due to the electrical charges binding the particles.

Substances are solids when the forces of attraction between the particles are strong enough to prevent the particles from moving freely. Solids have fixed shapes called lattices.

Liquids are fluid and can change their shape. They have flat surfaces and can take the shape of a container while maintaining a constant volume.

In a liquid, the forces of attraction between particles are too weak to hold them in a rigid formation. The particles glide past each other.

Substances exist as gases when the kinetic energy of their particles is high enough to completely overcome the forces that attract them. Gases, like liquids, can take the shape of a container. Unlike liquids, however, gases have enough kinetic energy to spread out and completely fill their containers.

The melting point of a substance is the temperature at which the kinetic energy of the substance's particles is just great enough to free them from the rigid lattice structure. The amount of energy needed to melt a solid depends on the strength of the attractive forces in the solid.

The freezing point of a substance is the temperature at which the kinetic energy of a substance's particles is reduced to the point that the substance forms a rigid structure. For example, liquid water freezes to form ice at 32° F.

The forces in iron, which melts at 2,765° F, are much greater than the forces in ice, which forms at 32° F.

A liquid boils when bubbles of vapour grow in the liquid, rise to the surface, and burst. The boiling point of a substance is the temperature at which the kinetic energy of the particles of that substance is great enough for them to completely escape the forces that pull the particles together. Each substance has its own boiling point. Water boils at 212° F to form steam. Liquid hydrogen boils at -436° F, and ethanol boils at 174° F. Some substances turn into a gas without passing through a liquid stage. This process is sublimation. Solid carbon dioxide sublimates from a solid to a gas at -173° F.

The above mentioned temperatures are based on sea level pressures of one atmosphere. Temperatures vary for higher pressures.

There are rules and regulations regarding how the elements form bonds based on the outer shell containing electrons. Some of these rules are as follows:

1) Each atom of an element contains a number of electrons that exactly equals the number of protons in its nucleus. The positive charges of the protons balance the negative charges of the electrons, and the atom has no overall charge. The electrons orbit the nucleus in layers called shells. There is a limit to the number of electrons that each shell can hold. The first shell, which is closest to the

nucleus, can hold up to two electrons. The second shell can hold eight electrons, and the third shell can hold eighteen.

2) The rows of the periodic table list elements in order of increasing atomic number, which is the number of protons in the atom of an element. Each row starts with an element that has only one electron in its outermost shell. At the end of each row is a noble gas, which has a full set of electrons in its outermost shell. The latter arrangement of electrons, which is called a configuration, is usually stable. Hence, noble gases seldom react, because their outer shells contain a full set of electrons. Elements react when bonds form between atoms as they gain, lose, or share electrons. As a result of these changes, each atom in a compound usually has a full outer shell of electrons. This is the stable electron configuration of the noble gas closest to each element in atomic number. The valency of an element is the number of bonds it must make to attain a noble gas configuration.
3) Metals usually have one or two electrons in their outer shells. They easily lose these electrons so that the next shell down becomes a complete outer shell. The non-metals at the far right of the periodic table are only one or two electrons short of a complete shell. Hence, they easily accept electrons from atoms of other elements. The valencies of these elements are the numbers of electrons they must gain or lose to form a complete outer shell.
4) Elements in the middle of the main block of the periodic table have outer shells that are three or four electrons short of a full shell. Carbon is an example of an element in the middle of the main block of the periodic table. Carbon has four electrons in its outer shell, which can contain a maximum of eight electrons. Carbon rarely accepts four electrons, because the negative charge of the ten total electrons would repel the electrons from the protons in the nucleus, making it difficult to form a compound. Instead, carbon atoms overlap their outer shells with shells of other atoms and share four electrons to make up the full count. The valency of carbon is 4.
5) Ionic compounds form when atoms of two or more elements trade electrons to form charged particles, or ions, of each element. The

ions have a noble gas configuration, and their charges balance each other to give the compound no overall charge. For example, in the formation of sodium chloride, sodium loses its eleventh electron to become a positively charged sodium ion, and at the same time chlorine gains an electron to become a chloride ion. The opposite charges of these two ions attract each other strongly. The ions bond together in a regular pattern called a crystal lattice. In a similar manner, the magnesium atom loses two electrons, and two chlorine atoms gain an electron each, to form magnesium chloride. Magnesium has a valency of 2. Ionic compounds form mainly between metals at the left of the periodic table and non-metals at the right of the periodic table.

6) Covalent bonding occurs when non-metal elements share outer-shell electrons to make up a complete shell. For example, in carbon dioxide, the valency of carbon is 4 and the valency of oxygen is 2. In carbon dioxide, one carbon atom shares one pair of electrons with each of two oxygen atoms. In this way, all three atoms fill their outer shells.

Hence, it can be seen that many of the 118 elements listed in the periodic table are atoms with outer-shell electrons that can join together to form many millions of different compounds with various properties. Strong forces of attraction called chemical bonds hold atoms together in these compounds.

In this manner, the Creator has transformed the dust into solar systems comprising planets and moons that contain molecules and compounds formed from the elements created in the cores of stars. Fortunately, here on Earth, the conditions are right for the dust to form seas and oceans, with millions of chemical compounds contained in the liquid water available in large quantities all over Earth's surface.

Heat and pressure are known to increase the rate of chemical reactions. Heat energy makes the atoms vibrate rapidly, making them reactive, while pressure forces the atoms together, making them more reactive. Both heat energy and pressure are present in the seas and oceans of Earth, together with currents and tides. Long ago, the elements in the water were forced to combine in every way possible, resulting in simple plants that could absorb energy and make food to support themselves in the water- and energy-rich environment. These plants had simple inputs, taking in carbon dioxide and sunlight to produce internal energy for growth and development, and releasing simple products, such as oxygen. From the water, these plants then spread to the land surface, where water was present underground and in rivers flowing across the land. Over millions of years, carbon atoms were lost from the atmosphere and oceans. These carbon atoms were stored in living and dead plants, while oxygen was put into Earth's atmosphere. Then the chemical compounds formed animal creatures to live off the plants and prevent their abnormal spread. These animal creatures left the crowded and polluted seas to come onto land, where plants thrived. Ecosystems and biomes formed where creatures lived in communities to survive.

It is interesting that many people belonging to a religious field would attribute such a splendid development to the work of a Creator, a Planner; and there are some valid reasons for this:

1. The water molecule can almost be considered a smart molecule because of what it can do. It can travel around the planet with the energy from the sun, and it can bring many elements and associated compounds to the oceans and seas, where millions of chemical reactions occur. It is not surprising that the sap of trees and plants consists of water, and that the blood of animals and human beings is composed mainly of water.
2. The nature of taking carbon out of the atmosphere with lifeforms, is crucial to keeping Earth's temperature stable, because carbon dioxide is a greenhouse gas that traps heat. Plants take in carbon dioxide and release oxygen into the atmosphere.
3. The process of emitting oxygen into the atmosphere with lifeforms is extremely important because it allowed human beings to inhale oxygen into their bodies, burn fires, cook food, communicate, become conscious, and become civilized with the ability to utilize energy.
4. The emission of oxygen into the atmosphere by lifeforms also helps to destroy small meteors by burning them up and blocks some harmful radiation, protecting Earth from disaster.

5. Earth's size enables it to trap an atmosphere approximately six thousand meters high, which is crucial for the weather and the plants and animals that live on the planet.
6. Lifeforms are composed of cells that have chromosomes and genes with specific instructions on how to make organs and all body parts to absorb and digest food and transfer energy to all parts of the lifeform. This enables the lifeform to live for a short period of time. No lifeform has detailed knowledge of internal processes of lifeforms. The whole operation is automatic; it just happens naturally.
7. The compounds making up life are complex, and the apparent programming whereby one egg cell and one sperm cell can fuse together to form a complex living organism that can live for a couple of years is remarkable, to say the least.

However, there are always the sceptics, such as atheists that argue that everything occurs through random processes with no plan. This is what human consciousness is. Human beings have the ability to choose; this is what separates us from everything else. The galaxies, the stars, the planets, the moons, the elements, the atoms, and the compounds all follow rules. The Black Holes must develop to form the galaxies. The stars must form in the nebulae within the galaxies. The stars must shine every day until they run out of fuel. The stars must produce the elements and make the energy that must flow out into the Space that surrounds them. The planets have their climates and atmospheres, and they must move in the field around their stars. The moons must orbit around their planets. Earth has seas with waves and life, where plants and animals live in the water and on the land. Only human beings can choose between what is considered right and wrong. It is not surprising that the early religious fields were all based on strict moral issues regarding good and evil. In the modern legal field, there are rules and laws that reward good and punish evil. One can be put in jail for committing a crime by the legal system. However, the fact remains that human beings can choose to commit crimes and break the rules deliberately for personal gain.

Through consciousness and the ability to choose, human beings have become civilized and educated. Through education, human beings have access to the Fields of Knowledge. By working in the Fields of

Knowledge, human beings can become like the stars, improving the fields, providing a driving force for fellow workers, developing ideas, and perhaps even creating new Fields of Knowledge for future generations of human beings.

5

LIFE

THE RULES THAT govern the transformation of inanimate life between states of solid, liquid, and gas is not complex. However, the rules for animate life are extremely complex, involving matter becoming alive in cells. Cells are very delicate structures that require suitable temperatures and pressures. Early forms of conscious life, such as fish, reptiles, and amphibians, were cold-blooded and could not regulate their own body temperatures. They had to rely on the energy of the sun to warm their bodies so that they could move around. Later forms of life, like the birds and mammals, were warm-blooded and could regulate their body temperature. This is important, as warm-blooded creatures can move around independent of the sun, making it easier to find food and water when it is cold.

Inanimate solids consist of atoms, molecules, or ions that are bonded together. The properties of a solid depend on the strength of the bonds that hold the solid together. Adding heat to the bonds holding the solid together provides kinetic energy to the solid. If the energy is sufficient to reach the melting point of the solid, the solid will melt and become a liquid. Adding heat to the liquid will add more energy to the bonds holding the liquid together. If the energy is sufficient to reach the liquid's boiling point, the atoms will escape from the surface as a gas. The reverse is also generally true. Removing heat from a gas can transform it back to its liquid state. Removing more heat energy from the liquid can transform the liquid to a solid. Elements and compounds are generally found as gases, liquids, and solids on the surfaces of planets and moons orbiting a star. On a star, matter exits as a plasma because conditions are too hot to form liquids and solids. Planets very close to a star are generally very hot, as is observed in our solar system, which means that Venus and Mercury are not suitable to support life. Planets very far from the star are generally too cold, as observed from Space Probes, which means that Mars, Jupiter, Saturn, Uranus, Neptune, and Pluto are not suitable to support life as we know it.

Lifeforms contain cells that have all the information needed to keep the organism alive and allow it to reproduce. These cells also carry out the processes necessary for life. The simplest living things are tiny single-celled organisms that appeared in the seas of Earth more than 3.8 billion years ago. The single-celled organisms have a cell membrane, or thin outer wall, that lets chemicals in and lets waste products out. Within the cell membrane is a jelly-like fluid called cytoplasm. The cytoplasm contains tiny structures, each of which has a function. The nucleus is the central structure of the cell. The nucleus contains the genes, which are like coded instructions to determine the cell's shape and function. Other structures within the cell release energy from food, deal with waste products, or protect the cell against attack from other organisms. Single-celled organisms are the most common form of life on Earth today and are known as monerans. Monerans can be seen only by using a very powerful microscope.

There are two main groups of monerans: bacteria and the plant like blue-green algae (cyanobacteria).

Single-celled creatures reproduce by splitting in two. Some bacteria can reproduce in fifteen minutes, which explains why they exist in large numbers.

Protists are also microscopic organisms; these include amoebas, tiny algae, organisms that feed like animals, organisms that trap the sun's energy like plants, and organisms that behave like both animals and plants. Amoebas reproduce by a complex process of cell division known as mitosis. Early life was extremely fragile because it had to have a delicate membrane to separate it from the outside world. Material had to be absorbed through the membrane to make the energy required for the organism to function, and waste products had to be released.

Earth's early oceans and seas must have been hot, turbulent places, because Earth is believed to have formed about 4.5 billion years ago. When Earth formed, it was a molten sphere. When these first organisms appeared, Earth had barely cooled and would have had a hot, thin crust and warm, turbulent oceans. Life is complex. The lifeform needs a membrane to separate it from the outside world. The lifeform within the membrane needs to take in material it must process to produce the energy needed to function and exist. Also, the lifeform within the membrane must eliminate waste products and fight off infection to be healthy.

What applies to these simple single-celled lifeforms applies to life today. However, over 3.5 billion years, life has grown and become more complex. The main difference between the single-celled organisms of early life and the multicellular organisms of today is that today's organisms have a brain and sensors that enable them to search for and find food and water for the production of energy. However, ever increasing population growth has resulted in competition for the available food and water, which in many cases has made survival difficult. Later in this chapter, I will present a full range of lifeforms, from the simple single-celled creatures to the complex multicellular mammals. Finally, human beings emerged from among the mammals to develop language and data storage, which were necessary to create the Fields of Knowledge.

It was only when human beings created the Fields of Knowledge that lifeforms could finally understand that they were created from the dust of the universe. For lifeforms on Earth, the dust becomes alive for a relatively short time to see the Creator's creation of billions of galaxies with billions of stars, all of which are creating the energies and elements that eventually result in life.

Earlier I described the water molecule as being almost a smart molecule. If this is the case, then the cell must be considered smart because of its ability to transform into the multicellular organisms that exist in the water, on land, and in the air. It is unbelievable that single-celled creatures can grow to be so complex on their own. If I asked a person from the religious field how single-celled creatures transformed themselves into complex organisms, he would say it was the work of the Creator and there is good reason for this:

1) To form multicellular organisms, the single-celled organisms need to organize themselves to be aware that there is more than one cell in the group.

2) Functions need to be assigned to the groups of cells to form the organs and other body parts.
3) Cells have to work out how to digest the food taken into their membranes to produce energy, which then has to be distributed to all other cells via a liquid transportation network.
4) Single-celled organisms have no protection from the raging seas, tides, and waves; however, multicellular organisms were able to grow shells to protect the cytoplasm.
5) Then there is the change in reproduction from one cell becoming two cells via a process known as mitosis to reproduction via meiosis. In meiosis, a male sperm cell is fused with a female egg cell; the process requires two separate entities to interact to create a new organism. This is a much more complex process, as it requires interaction between two separate organisms. It is useful for introducing variability and genetic diversity into the new offspring.
6) Further, completely new mechanisms had to be provided for organisms to leave the sea and live on the land and in the air.
7) The single-celled organisms were pushed along by ocean currents automatically. However, multicellular organisms living on the land and in the air had to propel themselves using wings, arms, and legs, which required a great deal of energy.
8) It is wonderful to see that organisms were provided with a brain and sensors not only to find their way around but also to find the food and water necessary to create the energy to be alive.

Thankfully, the mechanism by which cells operate has been worked out by human beings in the scientific field of biology. A living organism has trillions of cells. Different cells perform different functions. However, they all share the same structure. A plasma membrane separates each cell from its surroundings and allows material into and out of the cell. Inside each cell, tiny organelles float in a watery jelly-like substance called cytoplasm. The most important organelle is the nucleus, which is the cell's control centre. The nucleus contains genetic material in the form of deoxyribonucleic acid (DNA), which is composed of chromosomes containing sets of instructions called genes. Each gene instructs a cell to make a specific protein. Proteins build every cell in the bodies of plants and animals. Proteins also produce the substances or release the energy needed to make the cell work. Other organelles within the cell membrane include mitochondria, ribosomes, and the endoplasmic reticulum.

Food and water taken in by the lifeform has to be converted into energy for movement and nourishment for growth and the repair of cell tissue. An enormous number of chemical reactions have to take place within a living lifeform, involving molecules based on carbon. The sum of all the chemical processes in a living organism is called its metabolism. Metabolism has two parts. Catabolism includes all the reactions that break down large molecules with the release of energy. Anabolism is a process that uses simple molecules to synthesize proteins, fats, and other complex substances.

Cells reproduce by dividing in two ways. One way is mitosis, which occurs throughout the body and allows the body to grow and repair itself by replacing worn-out cells. The second way is meiosis, which occurs only in the testes of males and ovaries of females. Meiosis is a complex process whereby genetic information is passed on to the next generation using chromosomes found in sperm and egg cells. A person's genes are a combination of the genes of his or her parents. Genes occur in pairs—one from each parent. There are forty-six chromosomes in the nucleus of most human cells. For human beings, the process of cell division called meiosis makes sex cells that contain twenty-three chromosomes; there are twenty-three chromosomes in each egg of the female and twenty-three chromosomes in each sperm of the male. In the miracle of fertilization, when the sperm cell joins the egg cell, the simple chromosomes find their corresponding pairs to make a full complement of forty-six chromosomes. This becomes the blueprint for the new human being, complete with brain, sensors, organs, and operating systems. One pair of chromosomes, the sex chromosomes, differs from the twenty-two other pairs of chromosomes. The sex chromosomes carry genes that are not the same in both sexes. Males have a long X chromosome paired with a shorter Y chromosome. Females have two X chromosomes. The presence of the X and Y chromosome in the embryo causes male reproductive organs to form in the new lifeform. The presence of two X chromosomes in the embryo causes female reproductive organs to form in the new lifeform.

Interestingly, other animals have cells with different numbers of chromosomes. The cells reproduce by mitosis and meiosis, similarly to human cells. However, the genetic information in the chromosomes results in offspring of completely different sizes, characteristics, and shapes to human offspring. It is hard to believe that a few chromosomes can make such a variety of lifeforms.

Trees and flowering plants are unable to move like animals, so they use flowers and seeds as a means of fertilization to produce offspring. The unique look and taste of fruit on trees is an encouragement for insects and animals to satisfy their hunger and transport the seeds of the plants to new and perhaps more fertile areas, where they can grow and develop.

The whole process of transferring genetic data is extremely complex and requires programming of instructions in genes on a scale similar to the programming of microchips in modern computers. Yet genes and chromosomes come free, with their programs fully installed. The dance of the chromosomes in the fertilization process is remarkable, as is the formation of a complete lifeform. Though conscious, we and the other animals and plants are completely unaware of the processes involved. Only human beings know that this process occurs. Human beings and all the other lifeforms can all live to see the resulting creation for a relatively short period of time.

Needless to say, if I asked an atheist how single-celled creatures transformed themselves into complex organisms complete with fully programmed cells, he would say it occurred through random mutations and evolution. One has to make up one's own mind on this issue. It is interesting that only human beings can make up their minds on these and other issues, and this is described as human consciousness.

I had a dream. As I drifted out of consciousness, I could see the stars revolving from east to west. I dreamed that I was about to interview the Creator. I wondered what physical shape the Creator would have during the interview. I seemed to be waiting a long time, and I wondered if the Creator would come to the interview. My unconscious mind informed me that perhaps the Creator was the entire universe. I recalled hearing the words "God is everywhere" and "God is in heaven." Certainly, if God is everywhere and God is in heaven, then the Creator, or God, could very well be the universe. So I said, "Creator, what was your greatest rule in transforming the dust into consciousness?"

I waited for the response, wondering what frequency the Creator would use to reply. There was no outward response that I could detect. The response came, and surprisingly it was a consciousness-to-consciousness response. The response I heard can be converted to the following words: "My greatest rule in the Creation was fusion. I fused matter into extremely dense states in the Black Holes to create the fields of the galaxies. I fused atoms in the stars to create the elements and the energies to drive inanimate and animate matter on the planets and moons. I brought life to your world and fused sperm and egg to create organisms of life that led to human consciousness. Human consciousness has led to the Fields of Knowledge, where human beings can live on after the death of the body, as the stars do in the fields of the galaxies."

I said, "Creator, this is unbelievable." Unfortunately, the shock of getting a response awoke me, and from my window I could see that the stars were still revolving around the sky. I then realized it had all been a dream.

Clearly, the step of fusing egg and sperm to create a new lifeform was a major development. Single-celled organisms becoming two single-celled organisms by mitosis is an easy way of reproducing, because in this case one organism does not need to communicate with another lifeform. However, this process keeps the lifeform as a single-celled lifeform that is very similar, if not identical, to the original organism. Though one gets plenty of organisms through this process, there is not enough variation to develop things like language, writing, and painting.

The fusion of egg and sperm means that the Creator had to have two organisms that could communicate. The two organisms had to find each other, meet, and establish a bond. This would lead to language, communication, family units, groups, organizations, leaders, and even kings and queens. As the groups became more complex, there was an increase in data and the need for ever increasing data storage.

Once a person had language, that person would have to write things down so that data was not forgotten. This became extremely important as one became older and the memory cells began to decline. Once things were written down, it was important to store the data in a safe location so that it was easily accessible. Then, gradually, one would accumulate different types of data. In early times, one might have data like the following:

1. Location of good food
2. Names of famous people
3. Locations of good water
4. Places to hunt
5. Methods of cooking food, like recipes today
6. Ways to make fire
7. Plants that were safe and those that were dangerous
8. Animals that were safe and those that were dangerous

This would become like an early library for the people of that time.

So in this way, the rule of fusing egg and sperm, would eventually lead to language, data storage, and the Fields of Knowledge. It is amazing how things that appear simple can lead to major developments and to think that the Creator could have planned it.

There is only one organism that has access to the Fields of Knowledge and knows about it, and this organism is the human being, belonging to the class of creatures known as the mammals. Life has been around for approximately 3.5 billion years, and the fossil record seems to show that the Creator must have been very busy indeed. Living creatures take in energy from the sun, and after they die, some of this energy is retained in the dead remains. These plant and animal remains are buried in Earth's crust and subjected to varying temperatures and pressures as the land and water moves over long periods of time. Some of these plant and animal remains form beds of coal, oil, and natural gas, all buried underground. Through the Fields of Knowledge, human beings have been able to utilize this energy (discussed in chapter 7). The quantities of buried fossil fuels indicate that life has existed on Earth for a very long time. However, human beings with access to the Fields of Knowledge have existed for only about five thousand years. Some of the Lifeforms on Earth are as follows:

1. Single-celled organisms: Scientists estimate that more than three billion years ago, the first single celled organisms appeared in Earth's seas. These simple creatures are classified as monerans. These creatures are only visible through a powerful microscope. There are two main groups of monerans, namely, bacteria and plant like blue-green algae (Cyanobacteria)

2. Fungi, mushrooms, toadstools, yeasts, and slime moulds: Fungi do not have chlorophyll and cannot make food as plants do. Fungi make chemicals that rot the bodies of plants and other living things. The fungus gets nourishment from the decaying bodies of plants and the wood of trees. Mushrooms, toadstools, yeasts and slime moulds are classed as fungi. Some fungi are poisonous. However, some fungi are beneficial. The antibiotic penicillin is made from a mould, while bread is made from yeast
3. Lichens: A lichen consists of two living things in partnership or symbiosis. Single celled algae and fungi are lichens because the algae use photosynthesis to turn the energy from sunlight into food, while the fungus produces chemicals to rot and feed on part of the algae. The fungus in turn surrounds and protects the algae with its hard skin. Lichens grow on rocks, walls and the bark of trees. Lichens offer protection to the algae by being able to withstand extremely cold winter conditions. Lichens are eaten by animals such as caribou. Many fungi contain poisons which are harmful to animals. This is another form of protection for the algae living with the fungus.
4. Plants with roots, stems, and leaves
    a. Nonflowering plants: ferns, mosses, and liverworts, which have no flowers or seeds but disperse spores on air currents
    b. Flowering plants that produce seeds in the ovaries of their flowers and form fruit after fertilization to protect the seeds.
5. Trees: Trees have elaborate root systems to find underground water. Huge tree trunks take water up to leaves supported on tree branches, which have the ability to detect sunlight and grow strong enough to support the leaves in severe storms and harsh weather conditions. Like other plants, trees make their own food via a process known as photosynthesis. Water from the roots and carbon dioxide gas from the air combine to make glucose (a sugar) and oxygen gas. The trees use glucose as a fuel to make energy in a process called respiration. Glucose molecules are joined together into long chains. One chain, called cellulose, is used for growth and developing strength. Another chain, called starch,

is used as a reserve food store. Plants also make amino acids for proteins, enzymes, and hormones. Trees protect their seeds by encasing them in fruit. Trees do not choose where their seeds spread. Animals and insects disperse the seeds by eating the fruit or pollinating the flowers, as the case may be.

6. Marine invertebrates: These are the first multicellular animals, which formed prior to the development of backbones in nature, and they live in the seas and oceans. The oceans are still full of invertebrate animals. This group includes corals, sponges, jellyfish, starfish, and anemones.
7. Molluscs: These creatures have soft bodies covered by a mantle. Molluscs live in water and on land. This group includes clams, snails, slugs, octopuses, and squid.
8. Worms: These are legless invertebrates that live in the soil or in water.
9. Crustaceans: this group is known as arthropods and contains thirty-eight thousand species, many of which live in the oceans. This group includes woodlice, crabs, shrimps, prawns, and lobsters.
10. Spiders, centipedes and scorpions: These animals are arthropods, like crabs and insects. Spiders and scorpions are known as arachnids, and centipedes and millipedes are myriapods.
11. Insects: This group includes bees, flies, mosquitoes, wasps, dragonflies, grasshoppers, and butterflies. They can live anywhere on Earth and eat any food, but can only grow a few inches long. These creatures lay eggs and are capable of amazing transformation. The butterflies and moths go through a life cycle involving four stages: egg, larva, chrysalis, and adult. The caterpillar hatches from the egg and eats plants. When fully developed, it spins a silk cocoon known as a chrysalis, from which a fully formed butterfly emerges to fly away, pollinate flowers, find a partner, and repeat the cycle by laying eggs. Grasshoppers and cockroaches go through a cycle comprising three stages: egg, nymph, and adult. The nymph is a smaller version of the adult. However, it is unable to fly like the adult insect because its wings are not developed.
12. Fish: These animals were the first to have backbones and skeletons of bone; thus they are termed vertebrates. Fish have brains, eyes,

blood, spinal cords, and hearts. Fish take in water through the mouth, and water passes out via the gills. The gas exchange in fish is complementary to the trees. Fish take in oxygen and give out carbon dioxide. Fish are cold-blooded.

13. Amphibians: These animals are the smallest class of vertebrates. They were the first animals to colonize the land. However, they have to return to the water to breed. Most amphibians begin their lives in the water and breathe with gills. As they grow, they develop lungs and legs and are able to move on dry land. Amphibians are cold-blooded. These creatures include newts, salamanders, frogs, toads, and caecilians. Generally, females produce masses of eggs, or spawn, which are usually fertilized externally. After about ten days, the eggs hatch into tadpoles with gills. The tadpoles live and feed in the water. Eventually the tadpoles grow back legs, and then front legs, and then their tails shrink. Finally, after the adult frogs develop lungs, they clamber out of the water to find food on dry land.
14. Reptiles: These animals are cold-blooded animals that usually prefer to live in warm climates. They are characterized by their dry, scaly skin. Most reptiles lay leathery-shelled eggs on land. The shells prevent the embryos from drying out. Reptiles are believed to have been the dominant lifeform on Earth for 150 million years; the best known of these ancient animals are the dinosaurs. Today there are four main groups of reptiles:
    a. Alligators and crocodiles (about 25 species)
    b. Tortoises and turtles (about 250 species)
    c. Snakes (about 2,700 species)
    d. Lizards (about 3,700 species)
15. Birds: These animals are the largest group of warm-blooded vertebrates. All birds have feathers, beaks, and two front limbs that have been modified into wings. Most birds build nests of twigs and leaves on trees or in sheltered areas on the tops of mountains or cliffs. Male and female birds mate, and the female bird lays eggs in a nest. The male and female birds feed their young until they can fly away and find food on their own.

16. Mammals: These animals are warm-blooded animals that feed their young from milk-producing glands. Mammals are extremely diverse and can be found in almost every habitat on Earth. Mammals belong to the class Mammalia and consist of three main groups:
    a. Monotremes: These are animals that lay eggs. This group includes species like the platypus and echidna.
    b. Marsupials: These mammals, such as the kangaroo, give birth to partially developed young. As a result, the offspring stay in the mother's pouch and feed on her milk until fully developed.
    c. Placental mammals: This is a group of animals that gestate (develop young) inside their bodies, providing nutrients through the placenta in the uterus. These animals include flying mammals, such as bats; sea mammals, such as seals, dolphins, and whales; large herbivores or plant eaters, such as elephants and giraffes; powerful carnivores, such as dogs, cats, and bears; and primates, such as apes, monkeys, and human beings.

As Earth revolves around the sun with its axis tiled, various regions on Earth get differing amounts of energy. This causes variations in temperature and pressure on Earth. The escape of Earth's internal heat through volcanoes has caused changing mountainous landscapes. As Earth contains vast quantities of liquid water, various regions get differing amounts of rainfall. All this has resulted in varied climatic conditions. The various lifeforms of plant and animals above, show obvious preference for one type of climate over another. A biome is region of Earth characterized by climate and containing distinctive plant and animal life. The climate determines what types of vegetation are found in a region. In turn, the vegetation determines the types of animals found in the region. The generally accepted classification of biomes are tundra, taiga, temperate forest, shrub wood, tropical rain forest, grassland, desert, marine, freshwater, and estuary.

A biome is made up of smaller distinct regions called habitats. A habitat is defined as the area in which an organism lives. An earthworm's habitat is the soil. A conifer tree's habitat is the soil as well as the space above and around the tree. Organisms that live in a particular habitat are called a community. In any habitat, a number of species depend on one another in what are called mutually beneficial relationships. Plants provide food and shelter for animals. In return, animals help pollinate the plants. Plant-eating animals, in turn, are eaten by other animals. This is all part of the natural design. In fact, together, all the biomes make up the biosphere—the region of Earth that supports life.

Of all the millions creatures living in the biosphere, only human beings acknowledge the existence of a Creator. This fact helped human beings to grow from small clans into tribes and into large groups that established countries. Today a look at the world map shows the entire land mass covered by countries, which have their own rules and laws for the people within their boundaries. The biomes within a given country have become the natural resources of that country. Initially crops were grown and wells were dug into the ground to capture the underground water. However, as populations grew, these simple means of satisfying hunger and thirst had to be replaced by much larger farms and dams on rivers to capture and control large amounts of water. Also, sewage treatment plants had to be set up to treat wastewater before it entered the rivers and seas. All these issues led to the formation of cities in the various countries, where large groups of human beings and their chosen animals and plants could exist. The mere provision of food and water for a small group of human beings became the provision of food, water, housing, and energy for large groups of human beings in cities. Soon extensive trading occurred between various countries so that food and goods could be distributed, and this resulted in road networks, shipping networks, and aviation networks. Countries had to communicate with each other, and this resulted in vast communication networks. Today one human being can communicate with another human being anywhere on Earth via a simple radio device or computer.

Using large areas of land for human food production and housing has had some effect on the existing biomes. However, by creating cities where food, water, housing, and energy are readily available, human beings have come to dominate Earth. Human beings are the only creatures that have the consciousness to know of the existence of the biomes and the millions of other creatures that exist on Earth. Consciousness and the Fields of Knowledge reflect the responsibility human beings have in taking over what was primarily a natural process by which organisms existed in simple mutually beneficial relationships. Generally, organisms adapt to the natural conditions that exist around them. However, in cities, governments create rules and laws, and everything within a country has to abide by the rules and laws created by the government. This is certainly a new, and perhaps more interesting, way of living, provided the people making the rules and laws are honest and trustworthy. The Creator appears to have given human beings a great deal of power. Human beings have the power to kill an insect and remove a tree or any other organism almost on a whim. Human beings as a group can control the flow of water, control food production, communicate with other people anywhere on Earth, travel anywhere on Earth, and control vast amounts of energy. Consciousness and the Fields of Knowledge have given human beings control of life and a large portion of Earth.

6

CONSCIOUSNESS

OF ALL THE lifeforms that exist, only human beings are conscious of the Fields of Knowledge. Only human beings have an education system that teaches the skills of the Fields of Knowledge, so that people can spend thirty to forty years of their lives working and developing the Fields of Knowledge.

The Fields of Knowledge are so diverse that it is hard to believe that only one subspecies has access to it. The Fields of Knowledge improve the quality of life and provide a purpose and a value for life. Each Field of Knowledge is like a unique galaxy populated with human minds, developing and strengthening the field. In a particular field, new areas are developed, and sometimes old areas are dismantled as better theories and ideas arise. The fields are constantly changing and evolving over time.

A good example is the transportation field. People walked to get around or travelled on simple boats with sails. This limited the distance that could be travelled. Then, with improved knowledge, people invented the steam train and ship. The result was that people could travel farther faster. Then someone invented the bicycle. Then someone invented the motorcycle and car. This was followed by the airplane. Initially, the field of transport allowed people to travel a few kilometres using animal muscle power, but with the development of the Field to include airplanes and ships, people were able to travel around the world in a few days.

Some of the Fields of Knowledge are as follows:

- the transportation field
- the sports field
- the water industry field
- the air industry field
- the energy production field
- the food production field
- the clothes and grooming field
- the goods manufacturing field
- the building and construction field
- the mining field
- the religious field
- the medical field
- the communication field
- the spaceship Earth field
- the food preparation field
- the legal field
- the political field
- the business and economics field
- the music and entertainment field

A Field is a complex entity. Fields generally have a set of basic rules that define the Field. The rules and laws may be created by a singular mind or a group of minds. A field is held together by famous minds within the particular Field. Individuals who have passed away a long time ago may still influence a Field. In short, Fields are not limited to present individuals but are open to individuals from all past times. Obviously, there can be clashes between Fields. This is particularly seen in the religious, political, and legal Fields. A Christian Field, for example, can come into contact with a Muslim Field. Christians follow the broad teachings of Jesus and the Bible, whereas Muslims follow the broad teachings of Muhammad and the Koran. Strong radical elements within these Fields can cause massive turbulence and destruction. However, most normal individuals are only mildly affected as the two Fields settle down over a long period of time. In most cases, radical elements will exist in all Fields, resulting in some turbulence as the Fields come close to each other. In most cases, matters seem to settle after some destruction and devastation that effects a few individuals. The Fields are obviously held together by various famous individuals, which causes individuals within the Fields to have varying views. Some individuals might take extreme offence over minor disagreements, resulting in violence in certain elements of the Field. It is obvious that if such a situation gets out of hand or goes uncontrolled, a major turbulence can occur—perhaps even a major conflict or war. However, in all cases these events are like storms. The Fields settle after the storm. Some individuals are changed or transformed by the event, while others remain unaffected. Sometimes the event creates new famous individuals that strengthen the Field. Sometimes the event is a disaster, making the Field weaker and causing many individuals to leave—or even to be destroyed, in the worst case.

It is important to remember that the Fields provide a reason for living, as well as making it possible to live. The water industry field, the energy production field, the food processing field, and the building and construction field all make the quality of human life possible, and individuals within these Fields are provided with wages to enable them to live.

Fields are usually marked by events in time. Events can be something like a new famous individual being born in the Field or conversely a famous individual passing away. Consciousness is how the individuals within the Field react to the events. Some events might involve groups of individuals, such as World War I. Again, each thinking individual consciousness will be affected in a particular way by such an event. Obviously some individuals will be affected more than others.

The end result is that life is a complex web of Fields. The Fields appear to be constantly evolving in time. The Fields are held together by countless famous stars who came into existence as a result of events in time. Most normal individuals follow the lead of some famous individual and strengthen the Field. Major disastrous events cause major disruptions to the Fields, but everything settles down in time as individuals get used to the new conditions. In extreme cases, suicides occur when individuals are unable to get used to the new conditions, causing a massive disruption to their individual consciousness, making them believe that it is impossible for them to exist within their Fields.

Religious Fields are very complex. Christian principles of Jesus and the Bible form the basis of the Christian Field. Over time, new famous people arise and interpret the words and texts in a variety of ways, resulting in new rules or groups within the Christian Field. There are the Lutherans, derived from the teachings of Martin Luther during the Reformation. Another Christian group is the Anglicans, formed when the king decided to leave the Catholic Church, also during the Reformation. In this way, there is a spiralling effect out of the main Christian theology as new famous people create new subfields with slightly different rules and many individuals follow the new teachings. In the modern world, we have hundreds of new Christian groups with slightly different rules all claiming to be proclaiming the teachings of Jesus. This can make the whole Field more diverse and perhaps more interesting. It certainly provides people with many options. This Field has many institutions, such as seminaries, churches, and religious centres that provide support for the development of famous individuals within the Field. There always has to be an outpouring of knowledge within the Field by an individual to establish the Field and strengthen the Field so that it continues to exist among the millions of other Fields.

Sporting fields are generally less complex. Here one can see the rules that are defined by a Board. In the cricketing field, there are very definite rules for batsmen, bowlers, and fielders. An umpire ensures that the rules are followed during any cricket event. In addition, there are cricketing clubs that assist in the formation of cricketing Stars by having regular training sessions and forming the cricket teams. Then there are the events, the cricket matches between the clubs that make the cricketing stars. A cricket star is obviously a batsman, bowler or fielder that excels in many cricketing events. The Stars make the game interesting, generating the support of the public and hence holding the Field together.

Our consciousness gives us the ability to join the Fields of Knowledge. We have the ability to contribute to the many Fields, and if we are lucky, we might even become a star.

However, having the Fields of Knowledge is not sufficient. The knowledge in the various Fields has to be stored. In the past, knowledge was stored in paintings within caves, and then on slabs of stone. Having a language and communication skills is essential for storage of information. In the case of human beings, the invention of paper, and later the computer and software, were essential for storing data in books and electronically in databases on computer hard drives and memory sticks. Today human consciousness is stored in books or electronically in databases, where it is accessible via our wonderful machines known as computers.

Libraries are places where past human consciousness may be found and explored either in books or electronically via computers using a Search Engine.

For the purposes of this book, I visited a library to see if I could find some famous stars in the Field of science. Needless to say, it was an amazing experience. Even though these people had passed away, their contributions to the various Fields remain present in books or electronic formats so that they are available to people living in the present. Through books or electronic formats, the past stars have the ability to influence the new stars about to form, just as the past stars influenced me in writing this novel.

As the library was in the city of Sydney, in Australia, many of the people I came across had English backgrounds, and I realize I might have missed contact with many interesting people with non-English backgrounds. Life is like that; one must be grateful for the stars one comes across, for one is never going to come across all the billions of stars who have passed away or are alive today.

Hence, I am grateful to have come across the many famous people I have listed below. These are people who have contributed to the field of science over the generations, resulting in the explosion of Fields of Knowledge, improved machines, better technology, a higher standard of living, and the need to utilize energy. I have also included data received from some of the machines and telescopes sent out into space over the last fifty years.

Some of the past stars that I came across listed in time sequence are as follows:

- Archimedes (287 BC–212 BC): He was a famous mathematician and inventor. He invented the spiral pump for raising water, which is called the Archimedes screw pump.
- Nicolaus Copernicus (1473–1545): He studied mathematics, church law, medicine, and astronomy. After twenty years of work, he put forward a theory that Earth rotates daily on its axis and that Earth and other planets orbit the sun.

- Tycho Brahe (1546–1601): He proved that comets are heavenly bodies that orbit the sun.
- Galileo Galilei (1564–1642): He was an astronomer who improved the refracting telescope. He made many interesting discoveries using the telescope. Perhaps his most famous was the discovery of four moons orbiting Jupiter, showing that not everything orbits Earth.
- Blaise Pascal (1623–1662): He studied fluid pressure, showing that pressure in a fluid acts equally in all directions and that changes in pressure are transmitted instantly to the surrounding medium.
- Robert Boyle (1627–1691): He showed that the pressure and volume of a gas at a fixed temperature are inversely proportional. This is known as Boyle's law.
- Isaac Newton (1642–1727): He was a physicist and mathematician famous for his work in gravity. He developed the three laws of motion and discovered that white light is made up of multiple colours, similar to those displayed by a rainbow.
- Edmond Halley (1656–1742): He was a famous astronomer. He noted that three historic comets of 1531, 1607, and 1682 were so similar in characteristics that they had to be successive returns of the same comet, now known as Halley's Comet. He predicted the return of Halley's Comet in 1758.
- Benjamin Franklin (1706-1790): He was famous for his work in electricity. In a famous kite-flying experiment, he proved that lightning is a form of electricity.
- Joseph Priestly (1733–1804): He was a chemist and discovered oxygen in 1774. He also discovered that green plants give off oxygen and require energy from the sun as an input. He identified many gases, including ammonia, carbon monoxide, nitrous oxide, and sulphur dioxide.
- James Watt (1736–1819): He was a mechanical engineer famous for improving the efficiency of the steam engine which played a significant role for engines in the Industrial Revolution.
- Antoine Lavoisier (1743–1794) He founded modern chemistry, showing that air is a mixture of gases mainly composed of oxygen and nitrogen. He also proved that water contains hydrogen and oxygen.

- Alessandro Volta (1745–1827): He was a physicist who built the first battery cell in 1800. Volta's cell, also called the voltaic pile, was a stack of alternating plates of copper and zinc. Sheets of cardboard kept the metal plates apart, and the whole assembly was soaked in a solution of acid. The solution started the chemical reaction that produced electricity.
- John Dalton (1766–1844): He put forward the theory of atoms and proposed that molecules are made from atoms combined in simple ratios. He also published the first table of comparative atomic weights.
- Michael Faraday (1791–1867): A famous physicist and chemist. He discovered electromagnetic induction, which led to the invention of the dynamo and electric motor. His work contributed to the modern understanding of electricity, electrolysis, and the battery. He also showed that pressure can be used to turn a gas into a liquid.
- Charles Darwin (1809–1892): He is famous for his theory of evolution, which postulates that species were not created individually but developed over long periods of time, influenced by their surroundings and other species.
- Gregor Mendel (1822–1884): He was a priest. While researching inheritance in plants using edible peas, he discovered and developed the principles that govern genetics. These include the law that traits are inherited independently of each other through genes and chromosomes in cells.
- Louis Pasteur (1822–1895): He was a chemist. He showed that microbes cause fermentation and disease. He developed the process of using heat to kill germs (pasteurization) and popularized the sterilization of medical equipment, which helped save lives.
- Dmitri Mendeleev (1834–1907): He was a famous chemist who formulated the periodic law, creating a farsighted version of the periodic table of elements. He used his version of the periodic table to correct the properties of some already discovered elements. He was also able to predict the properties of eight elements yet to be discovered.
- James Clerk Maxwell (1831–1879): A famous physicist, he wrote down the laws of magnetism and electricity in mathematical form.

- Thomas Alva Edison (1847–1931): He was a famous inventor and manufacturer. One of his famous inventions was the electric light bulb, which was used to replace gas lamps. He also invented the carbon transmitter for telephones, which improved the audio system of the telephone.
- Sir Charles Algernon Parsons (1854–1931): he was the inventor of the steam turbine, which was used for driving electric generators and propelling ships.
- Joseph John Thomson (1856–1940): A famous physicist, his work led to the discovery of the electron.
- Henry Ford (1863–1947): A famous car designer and manufacturer. He introduced revolutionary new mass production methods, including large-scale production plants using standardized interchangeable parts. He also introduced the first moving assembly line for manufacturing motor vehicles.
- Marie Curie (1867–1934): She discovered radioactivity. She isolated the elements polonium and radium and discovered the element plutonium.
- Ernest Rutherford (1871–1937): A famous physicist who worked with J. J. Thomson at Cambridge University, he discovered the presence of nuclei in atoms.
- Guglielmo Marconi (1874–1937): He invented wireless transmission of data and broadcast messages using radiotelegraphy in 1907.
- Albert Einstein (1879–1955): A famous physicist, he related matter and energy, stating that matter and energy are interchangeable, via the famous equation $E = mc^2$. This equation implies that a very small atomic mass can be converted into an extremely large amount of energy. This mass-to-energy conversion is what makes the stars shine for billions of years.
- Dorothy Hodgkin (1910–1994): She used computers in the field of X-ray crystallography. She discovered the atomic structure of vitamin B12 and other complex molecules, such as penicillin and insulin.
- The astronauts visited Earth's moon from 1968 to 1973: The astronauts reported that the dust on the surface of the moon had not turned into consciousness that they could detect. In fact, the dust

was a major problem, because it got onto their spacesuits and made the lunar module very dusty. This is a very strong indication that existence in a Goldilocks Zone might not be the only factor leading to the development of consciousness on a planet or moon. The moon landing did show that human beings could live on other worlds provided they had equipment and systems to shield them against all possible dangers, including the lack of oxygen in the air. The astronauts visiting the moon had spacesuits with extensive life-support systems. The cost of one spacesuit was estimated at $2 million US.

Machines have also made some contributions in investigating space and other planets.

- The Hubble Space Telescope was placed in orbit five hundred kilometres above Earth. It provided clear, undistorted images of other galaxies in the universe. Looking at what appeared to be an empty patch of sky, Hubble saw many galaxies, showing that the universe contains billions of evenly distributed galaxies. Located above Earth's atmosphere, Hubble had an unobstructed view of the universe. (Lighting and Earth's atmosphere are known to distort optical telescopic observations from Earth.)
- Other telescopes, including radio telescopes, microwave telescopes, optical telescopes, ultraviolet telescopes, and X-ray telescopes, have been used to study the universe.
- Thousands of satellites orbit Earth, providing information on the weather, news, geography, global communications, and global positioning systems (GPS).
- Machines visiting Venus sent picture data indicating that the dust on the planet's surface was very hot, with very high surface pressures and a poisonous atmosphere. This was probably caused by a runaway greenhouse effect, by which the atmosphere traps sunlight and heat, preventing them from escaping into Space. The images showed that the surface rocks are so hot that there are rivers of molten rock flowing on the surface of Venus.
- Machines visiting Mars sent picture data indicating that the dust was very cold and dry. However, some pictures indicated that

water might have existed on its surface billions of years ago when the sun was young and more energetic.

- Machines visiting and flying by the outer solar system sent back pictures indicating that the dust on the surface of the outer planets was probably gaseous, with high-velocity winds and possible storms. However, the outer planets had many more moons than previously seen from Earth-based telescopes. All the moons were unique, with beautiful surface features and climates. Some moons seemed to have liquid water in their interiors. However, being far from the sun, they would need their own sources of energy to support life. Saturn's beautiful system of rings turned out to be blocks of water ice ranging from the size of sand particles to the size of large houses.
- Generally, no lifeforms were discovered in all the space observations.
- However, the Cassini orbiter visiting the Saturn system sent a probe called Huygens to land on the surface of Saturn's orange-coloured moon, Titan. Scientists interpreting the data from Cassini and Huygens have determined that Titan contains seas and lakes of liquid methane. In the cold temperatures on Titan's surface, methane, a gas on Earth, is a liquid, and water, a liquid on Earth, is a solid. Scientists claim that methane falls from the sky of Titan as rain, and rivers of methane flow through masses of solid ice on Titan's surface, forming the lakes and seas. If this is true, water might not be the only liquid on the surface of a planet or moon. In fact, any compounds might exist as liquids on other planets and moons, depending on the surface temperature and pressure. This is significant, because it could give the Creator all manner of options for liquids on the surfaces of planets and moons. This could lead to all manner of lifeforms being present on at least a few of the billions of moons and planets in the universe.

From the work of the above and other famous people in their Fields, there was an explosion of inventions and discoveries in many Fields of Knowledge, resulting in the modern world using all manner of machines, including cars, ships, trains, buses, farm machines, food processing machines, factory machines, planes, helicopters, jet engines, rockets, satellites, microchips, computers, printers, cameras, film projectors, robots, mobile phones, telephones, pumps, televisions, radios, telescopes, microscopes, electric shavers, electric toothbrushes, air conditioners, fluorescent luminaires, light-emitting diode (LED) luminaires, and all manner of medical equipment.

Everyone knows that all lifeforms require food and water inputs. Then the bodies of lifeforms have organs that convert the food into energy. Lifeforms also have a transportation system that uses a medium like sap or blood to transport energy around to all the cells so they can function according to their genetic code. Also, all lifeforms emit waste products, such as oxygen in the case of trees and urine and faeces in the case of animals.

Machines are far less complex because they do not have cells all over their bodies. However, to do work, machines have to use some form of fuel as an input, and there is some waste output. Some machines need elaborate cooling systems and noise-reduction systems. As the number of machines increases, the fuels used to produce energy becomes increasingly important, as does the disposal of waste products from the use of fuels.

Investigation of the universe is very important for human consciousness. If there are billions of galaxies with billions of stars in each galaxy, then one has to wonder if the Creator made living material elsewhere in the universe. Why would billions of planets and moons exist, many of them unobservable from Earth by human beings? We know that all the planets and moons are unique, with varying temperatures, pressures, climates, and elements. In addition, they can be subjected to varying gravitational forces from their stars and from planets, moons, asteroids, meteors, and debris around them. All the above factors indicate that the dust on planets and moons of other stars can have all sorts of constructive and destructive influences. It is remarkable that we live on a planet with an escape velocity of only eleven kilometres per second. It is also remarkable that human beings have invented machines that can easily reach Earth's escape velocity to travel to other galaxies. Perhaps it is destined for humanity to take consciousness to the other galaxies.

Maybe, although Earth is not in the centre of the universe geometrically, it is the centre for consciousness. Because, the whole universe appears to be observed from Earth via complicated machines like telescopes and Space Probes.

Just as matter has to reach the escape velocity of a star to bring energy to the planets and moons, our machines are capable of reaching the escape velocity of Earth to bring consciousness to the Milky Way Galaxy and perhaps other galaxies.

7

ENERGY UTILIZATION

THE UNIVERSE CONTAINS billions of galaxies. Each galaxy contains billions of stars orbiting around its central Black Hole. The stars fuse hydrogen in their cores to form helium and energy. The heat generated makes a star's outer layers a plasma. Energetic particles reach the star's escape velocity and leave its surface at approximately three hundred thousand kilometres per second. The Fields surrounding the star cause these particles to emerge at various frequencies or energies as follows, listed in order of increasing energy:

- radio waves—low frequency
- microwaves
- infrared waves—heat
- visible light: red, orange, yellow, green, blue, indigo, and violet
- ultraviolet
- X-rays
- gamma rays—high frequency; very energetic

The star's energy falls on the surfaces of the planets, moons, and other objects orbiting the star. In this way, through the utilization of the energy from the star, the objects orbiting the star acquire energy and have varying temperatures and climates. Lifeforms existing on planets or moons must be able to utilize the energy from the star, or the internal heat energy of the planet or moon, in order to live and grow.

On Earth, plants utilize the energy of the sun by having leaves containing chlorophyll, which enables plants to make their own food via a process called photosynthesis. Water from the plant roots and carbon dioxide gas from the air combine to make glucose (a sugar) and oxygen gas. The plant uses glucose as fuel to make energy in a process called respiration. Glucose molecules form long chains called cellulose used for growth and development of the plant. Plants also make starch and amino acids for protein, enzymes, and hormones. In this way, the energy of the sun is converted to energy that drives the plants and trees. Plants do not have brains, because they can find energy without needing to travel. However, they are able to sense light and temperature, and their branches automatically respond to enable them to get the most energy while remaining fixed to the ground with an elaborate system of roots.

Earth is able to utilize the sun's energy by having huge liquid oceans that cover 70 per cent of its surface. The sun's energy and Earth's internal energy prevent the oceans from being frozen solid, and this provides a readily available supply of liquid water for plants and animals.

Most people put clothes on a clothesline to dry in summer. Within a couple of hours, the clothes are dry. The sun's energy is utilized in turning the liquid water in the clothes to water vapour, which escapes into the atmosphere. This rule applies to all the liquid water on Earth's surface. The atmosphere is constantly being saturated with water vapour as the sun's energy converts liquid water into vapour. The vapour in the atmosphere cannot reach Earth's escape velocity, and Earth's gravity eventually brings the vapour down as rain. This creates the thousands of rivers that flow across the land back to oceans, bringing water and nutrients from the soil to the oceans. All this is essential for the growth of plants and trees in the oceans and on the land. Plants and trees that bear fruit, vegetables, and leaves are sources of food for hungry animals.

In order to find food supplied by the plants and trees, hungry animals need to have a brain and sensors. Fruits and vegetables are essentially sources of energy created by the sun through the trees and plants. Lifeforms, such as fish in the sea, birds in the air, and insects and animals on the land, all possess brains and sensors. These lifeforms need to move great distances for feeding, reproduction, and caring for the young. All these tasks require a lot of energy. Herbivores obtain energy by feeding on the plants. Carnivores obtain energy by feeding on the herbivores. Omnivores obtain energy by feeding on the herbivores and carnivores. In short, the sun's energy is utilized by all animals to live in their biomes.

The sun's energy distribution around Earth is not uniform. Hence, the utilization of the sun's energy varies greatly across the surface of Earth. Some regions get very little rain, while other regions receive plenty. Some regions are dry deserts, tropical areas are generally hot and wet, and polar regions are cold. These climatic conditions generally determine which plants and animals can survive in a particular region. A biome is a plant and animal community that covers a large geographical area.

Hence, deserts are very dry regions where few plants grow. They may be hot or cold.

Grasslands are most common in temperate regions. In tropical regions with long dry seasons, the typical grassland is savanna with scattered clumps of trees.

Scrublands are areas where bushy forms of vegetation dominate. Summers are hot and dry, and fires are frequent.

Taigas, also called boreal forests, are regions of subarctic coniferous forests. Winters are cold and long.

Temperate forests are found between tropical and polar regions; the climate here is mild with moderate rainfall.

Tropical rain forests grow where the weather is hot and humid all year. They form the richest biome in terms of the variety of plants and animal species.

Tundras are cold, dry regions where the subsoil is permanently frozen.

Oceans are low regions. Rivers generally drain into the lakes and oceans. Rivers bring nutrients, dissolved rock, and minerals into the oceans. Hence, oceans and lakes provide shelter to a wide variety of plants and animals. The energy from the sun and Earth's internal energy keep the water in a liquid state. The general buoyancy of the water helps single-celled creatures, small creatures, and fish to easily float and swim around. The nutrients are constantly restored and refreshed by the rivers, which flow into the lakes and oceans, providing food for the many lifeforms that exist there.

The above are natural biomes because the energy distribution of the sun is natural. The plants and animals have to find their own food and water in the various biomes described above.

Energy can be effectively utilized by transforming energy from one form to another. Human beings have been able to work out the non-renewable and renewable forms of the sun's energy. This has led to the development of the city and town as the new human biome.

Human beings have created cities and towns, where the energy has been derived through the Fields of Knowledge. Cities and towns are places where food and water are readily available. Generally, cities have houses for residences and offices, streets for access, transportation networks, power networks, and communication networks. This makes the cities relatively comfortable places. The ever increasing demand for cities has had some effect on the natural biomes, where many other species of plants and animals find themselves in protected parks and reserves—or zoos, in some cases,

It is generally accepted that human beings came to understand energy through fire. Fires formed naturally during periods of excessive heat and during lightning storms. The flames of fire radiated heat energy. The heat energy from fire provided warmth, lighting at night, protection against larger aggressive animals, and a means of cooking food by killing the bacteria that seemed to be everywhere. With the ability to create fire by friction, human beings began the long journey of civilization that would lead to storing data in books and then computers. Effective data storage made reliable data and education available to all human beings. With education came access to the Fields of Knowledge. Through the Fields of Knowledge, human beings are able to transform the energy of the sun to do all manner of work to develop the ever growing cities and towns. The world map has been divided into countries and states that are controlled by governments organized by human beings.

The sun's energy has been driving Earth's natural biomes for many millions of years. The result is large coal, oil, and natural gas deposits that have been buried underground for many millions of years. Coal is an impure form of carbon that formed from the remains of prehistoric plants that were buried and compressed over long periods of time. With compression underground, high temperatures and pressures broke down the plant remains to form coal. Coal consists of 98 per cent carbon, together with hydrogen, nitrogen, and sulphur. Coal contains stored energy and burns easily to release large amounts of heat. Coal was the main fuel used in the Industrial Revolution at the end of the seventeenth century.

Coal is burned in power stations to generate electricity, which is supplied to houses and offices in towns and cities. In this way, the sun's stored energy is converted into electricity for lighting and driving all household and office appliances, including refrigerators, stoves, air conditioners, and computers.

Oil is formed from the decomposed remains of tiny organisms that lived in the seas and oceans many millions of years ago. When the tiny organisms living in the ancient seas and oceans died, they were covered with layer upon layer of sediment that settled in the water as a result of Earth's gravity. Over time, the weight of those layers of sediment turned the organic remains of the tiny organisms into crude oil. The appearance of crude oil varies from a pale yellow liquid to sticky black tar. More than 50 per cent of the world's known reserves of crude oil are in the Middle East. Crude oil, or petroleum, is a mixture of chemical compounds that consists mainly of hydrogen and carbon. This mixture is separated and treated in oil refineries to produce a huge range of petrochemicals or oil derivatives. The main derivatives of oil are fuels, such as diesel, jet fuel, and petrol. In this way, the sun's stored energy is transformed into fuels that drive motor vehicles and aeroplanes. The burning process produces hot gases that expand and force pistons in engines to move. In this way, chemical energy in the fuel is converted to heat energy and then to mechanical energy to drive wheels and other associated parts, resulting in work being done.

Natural gas, like crude oil, formed over millions of years from the remains of marine organisms. Natural gas is found with or near oil reserves. The main component of natural gas is methane, the simplest hydrocarbon, whose chemical formula is CH_4. Mixed with methane are smaller amounts of other hydrocarbon gases, such as ethane, butane, and propane. Natural gas is piped to houses and is commonly used as a fuel for heating and cooking. Natural gas provides about 20 per cent of the world's energy supply.

Coal, oil, and natural gas are examples of fossil fuels because they contain energy from sunlight that was trapped by organisms that lived millions of years ago. Fossil fuels store energy in the form of chemical energy. Fossil fuels are non-renewable fuels. Plants and trees remove carbon from the atmosphere; hence, when they are burned to release energy, they emit carbon back into the atmosphere. This is not a natural process; there is a certain amount of pollution caused by power stations burning coal and vehicles burning petrol or diesel.

About ten per cent of the world's electricity demand is generated from nuclear energy. In a nuclear fission reactor, atoms of uranium or plutonium produce heat energy as they split into smaller atoms. That heat is used to boil water, just as heat from furnaces is used in a conventional coal power station. Uranium atoms are believed to have formed in supernova explosions. The fact that those heavy elements are found on Earth's surface is a marvellous example of planning and recycling that allows for these products to be retained for use by a species wishing to generate vast amounts of energy to make life more enjoyable and useful. This may very well be one of the important features of the creation process. One of the drawbacks of nuclear fuels is that the materials generated by nuclear energy remain dangerously radioactive for thousands of years after they have been used in reactors.

Electrochemical cells and batteries were invented in 1800. They are a self-contained and often used as a mobile source of energy. The chemical reactions that operate within the battery cell occur in two parts or half-reactions. One half-reaction produces electrons at the negative electrode, or anode. The other half-reaction consumes electrons at the positive electrode, or cathode. A solution called an electrolyte is in contact with the electrodes and provides the materials for the reactions. Batteries have a positive terminal and a negative terminal, with a voltage drop of 1.5 volts, 3 volts, 6 volts, or 12 volts being typical. When a circuit is connected across the positive and negative terminals of a battery, a current flows from the negative terminal to the positive terminal, utilizing the chemical energy stored in the battery.

There are two types of battery cells. In primary cells, electricity or current stops flowing when the chemical reaction is complete, and the cell must be discarded. Secondary cells are able to be recharged by passing a small amount of current into the cell. This reverses the chemical reaction, and the cell can be reused after a few hours of charging. Lead-acid secondary-cell batteries are used in road vehicles to power the vehicles' electrical systems. While the vehicle is operating, the engine drives a small generator that charges the battery. Rechargeable batteries are also used extensively with mobile phones and laptop computers.

Batteries are an extremely important means of powering machines when fixed power is unavailable. Unfortunately, batteries do not last forever; their lives are dependent on the chemical reactions that drive the battery cells. Even rechargeable batteries have to be replaced after a number of recharges. The proper disposal and recycling of batteries are necessary to keep the environment safe.

Perhaps one of the greatest achievements of human beings is the use of chemical energy to blast rockets, satellites, and telescopes into orbit around Earth. Here heat energy blasting out of the rocket in a carefully controlled explosion can drive the rocket upward to attain Earth's escape velocity. Free from Earth's gravity, the rocket floats in space as it circles Earth. The use of energy has freed human beings from being constrained to Earth. According to scientific evidence, it was a mere 4 billion years ago when multicellular creatures began to swim and pulsate in the buoyancy of Earth's oceans. It is a great achievement to be capable of floating in Space above Earth. However, the human body is composed of muscles and a brain adapted for life on Earth. This might limit human achievements in Space and make robots more suitable to adapt to ever changing conditions, particularly on the moons and planets of star systems other than the sun.

Fossil fuels and nuclear fuels are non-renewable sources of energy and will run out some time in the future.

There are, however, renewable sources of energy that operate directly off the sun, and these will last as long as the sun shines in a stable equilibrium state, converting hydrogen to helium. These sources of energy do not produce pollutants and are therefore better for the environment.

Hydroelectric power stations use flowing river water as a renewable resource to do work by converting the kinetic energy of water into the mechanical energy of a spinning turbine. The rotating turbine is placed in a magnetic field, and this generates electrical power, which is transferred to cities or towns via copper conductors and electrical substations. Rivers are usually dammed to control the flow of water, so turbines placed at the outlets are a bonus, as they provide electricity. However, there are always the installation costs and the maintenance costs of the electrical equipment. Also, electricity cannot be generated during periods of drought, when the river levels are extremely low. Nevertheless, the sun is still the primary source for this type of power generation. The sun's radiation generates the heat that transfers energy to liquid water, making it a vapour. The atmosphere is saturated with water vapour, and as the vapour rises when it crosses high mountainous land, it cools, losing energy and eventually falling as rain into the river. The river flows because the high mountainous regions are higher than the oceans and Earth's gravity exerts a force on the liquid water. At the turbine, the operator controls the water speed by adjusting the turbine blades, and the kinetic energy of the water is converted to electricity. This is a very clean form of generating electricity. However, alternative backup systems must be available for use during periods of drought. This is where coal power stations have a significant advantage. Because the supply of coal is readily available, it does not require an elaborate backup system. The main problem with coal power stations is a long-term problem caused by the release of carbon as an exhaust gas and a limited supply of coal, which will run out after hundreds of years at the current rate of use. Obviously, the use of coal was preferred early on, because of the economics and because it's polluting effect was not known, as the Industrial Revolution had not occurred before.

Another renewable source of energy driven by the sun is wind power. The sun heats Earth's atmosphere by differing amounts, resulting in varying temperatures. Also Earth's rotation, its tilted axis, and its water distribution result in widely differing temperatures and pressures across its surface. According to the laws of thermodynamics, heat energy travels from areas of high heat to areas of low heat. This causes wind. Wind is sufficient to drive ships with sails. Wind was used in ancient times to drive windmills that transferred the energy of the wind to the mechanical energy of turning wheels to grind maize or wheat. Earth's polar regions are generally cold, while the regions between the Tropic of Cancer and the Tropic of Capricorn are generally warm. Hence, some winds are reasonably consistent, such as the Trade Winds. However, climatic conditions can change drastically, and in modern times sailing ships have been replaced with ships using nuclear, steam, or electrical power.

Today windmills pump water from underground sources and also drive turbine generators to make electricity without pollution. Wind machines have two or three blades up to fifty metres long, arranged as on the propeller of an aircraft; these are referred to as wind turbines. The wind turbines drive electric generators located inside the heads of the turbines, which stand on towers over 100 metres high. Computer-controlled motors alter the angles of the blades and the directions they face in order to suit the speed and direction of the wind. In a moderate wind, a turbine generates three hundred kilowatts of power. To be effective, a number of wind turbines can be grouped together in a wind farm. Wind turbines need a minimum wind speed of twenty kilometres per hour to start turning. By connecting the wind turbines to a grid system, the lack of wind in one place may be made up for by wind blowing elsewhere.

Another renewable source of energy driven by the sun is solar power. Solar cells are assembled into solar panels, which are used to convert solar energy into electricity directly. However, an electric battery system must be used to store the energy, because the sun generally produces only six hours' worth of usable electricity per day. The sun rises in the east and sets in the west. When the sun rises and sets, its energy is relatively weak, and at night there is no sunlight. This variation means that additional storage capacity is required to provide the necessary supply at all times.

Sea power is another renewable energy resource. Oceans absorb and store heat energy from the air. This heat drives winds across the surfaces of the oceans, causing waves to build up. Devices have been invented that can change the up and down motion of waves into a rotational motion that can drive a generator to produce electrical energy. Also, the gravitational pull of the sun and moon cause tides that ebb and flow each day. Devices have been invented that can cause the sea water to flow through turbines as the tide rises and falls, converting the energy into mechanical rotation of the turbines, which generates electrical energy. Sea power is relatively expensive and can have some effect on fish and marine lifeforms.

All renewable sources of energy require some backup support to make them reliable. Perhaps the only way solar power and wind power might be effective is to operate them on a worldwide basis. A system needs to be designed whereby solar panels and wind farms are located all around the world—particularly in the tropical zones. All the solar power stations and wind farms would feed into a common grid or power network. In this way, the night-time load could be supplied by the sun from the day side of the planet, and when the winds are weak in one area, the solar power and wind power from another area could supply the load. This worldwide system would utilize the solar and wind power far better, because the sun is always shining somewhere and is directly overhead somewhere on Earth. This system would require countries to cooperate on a global basis, which might require a lot of rules and regulations. However, the end result would be cleaner energy for cities all over the world. Such a system might become necessary to reduce pollution when Earth's supply of fossil fuels runs out.

Detailed elaborate calculations are not a part of this book. However, it is easy to see the feasibility of a worldwide system from earlier writings in the book—namely, that 70 per cent of Earth's surface is covered with liquid water or solid ice. The sun's energy is able to vaporize substantial quantities of water and ice, which eventually falls to Earth as rain or snow. This system, together with Earth's gravity, powers thousands of rivers and streams all over the land surface of Earth. The sun's energy drives the wind system all over Earth's surface. Excess energy from the sun results in severe heat conditions, floods, and storms. It is therefore easy to see that solar power stations and wind power stations located in strategic locations all over Earth's surface, with a concentration in the equatorial regions, together with computer monitoring systems with interconnected switchboards, can easily connect and power the entire Earth with clean electrical energy twenty-four hours a day. Technically, some of the solar power stations or wind power stations should be active as the sun shines overhead on them or the wind drives them. With experience, more power stations could be added to the worldwide network until it was able to operate more efficiently.

One might ask, "How does energy utilization connect to the journey from dust to consciousness?"

To reach the higher levels of consciousness—and, in particular, human consciousness—one needs to utilize energy on a large scale to support the new biome of the city and town. Cities and towns provide easy access to food, water, shelter, transport, lighting, communication, and work for millions of human beings. Hence, the new biome needs a great deal of supporting infrastructure and resources as described below.

Food, consisting of plants and associated fruits and vegetables, has to be grown in vast quantities to sustain the people living in the towns and cities. Similarly, the food consisting of animal products has to be grown on large farms to sustain the people living the cities and towns. Water services have to be provided to cities and towns, and the sewage or wastewater has to be carefully removed and treated so diseases and plagues do not destroy the populations in the towns and cities. All these services require energy utilization, which is achieved through the ever increasing Fields of Knowledge. Through an intensive education system, many stars are formed that are able to sustain and develop the Fields of Knowledge, which are able to support and develop the cities and towns where human beings are conscious.

I stated in the previous chapter that human beings are each composed of trillions of cells with chromosomes and genes fully programmed to take in food and water and convert these inputs into energy. The energy is transported by the bloodstream to all the living cells in the body so that a human being can live a healthy life. This means human beings need to consume vast quantities of plant and animal material to live healthy lives. Obviously, something is needed that can do work without consuming large quantities of plants and animals; this would greatly improve the journey from dust to consciousness.

The end result is the modern world and an entirely new species—the machine. The machine lives directly off energy and does not require plant and animal matter as a source of energy. In the original biomes, the energy generated from the sun was converted by plants into food, and animals ate plants to get energy for work via muscle power. In cities and towns, energy is used by machines to do work.

Through work in the Fields of Knowledge, human beings learned that machines provide an efficient way of using energy to do work. Machines are used to build cities, roads, tunnels, and transportation networks. Machines are used in the vast communication networks that link the entire Earth. Machines are used to construct and maintain houses and workplaces. The waterways are full of ships and sailing vessels. The air always has aircraft flying around in it. The land is full of cars, trains, and trams. Human consciousness has created machines to transform energy to do work that is necessary for human beings to survive in cities and towns, where clean water, food, and services are easily available. We even need machines to help us leave the Earth's surface and travel to a space station. We need a rocket to launch a satellite for a communication system. We use various forms of lights to illuminate our houses, workplaces, roadways, and public spaces. Water pumps are used to pump water around not only for drinking but also for growing crops. Machines monitor and control the energy delivered and issue alarms and warnings to alert operators. Machines are used in food production and in the food-processing industry. Machines can be used to plough the land, sow seeds, reap harvests, and pack goods in boxes, bottles, or cans. The car industry uses machines to do painting, welding, and some assembly work. Factories are full of exotic machines all busily working without sleeping, eating, carrying out industrial action, or collecting wages. The journey from dust to consciousness has led us to a world with cities and towns, in which machines are the most efficient and reliable means of converting energy to do the required work.

Machines are not born; they have to be manufactured to design specifications. Machines must be tested after manufacture to ensure that they comply with the specifications. This provides a safety standard for the use of machines in order to ensure that they do not cause any malicious damage.

If the Creator loves us, then surely this is manifested in our consciousness, which allows us to create the Fields that give rise to the human stars, which help to create the numerous and magnificent machines all around us.

Unfortunately, the energy that drives the machines are mainly derived from fossil fuels, with minimal use of solar, wind, and water power. This results in some pollution of the air over long periods of time. In only three hundred years of usage since the Industrial Revolution, not only have the fossil fuels become difficult to find on Earth, but the carbon released seems to be effecting the climate. Obviously, trees and plants removed carbon from the atmosphere in the past, so it is not known what damage putting large quantities of carbon back into the atmosphere will do. In any case, at the current rate of usage, it is expected that the fossil fuels may last a couple of thousand years at the present rate of consumption. However, as human beings are expecting to live as a species on Earth for at least a few million years, alternative means of sustaining cities and their machines will be required in the future. Hence, efficient renewable sources of energy will certainly be required in the future to sustain the towns and cities.

People living in large numbers in cities and towns results in some water pollution. When humans eat food for energy, they excrete waste products, which results in some water pollution. Even though water treatment is practised in first- and second-world countries, it is not a perfect process.

Some contaminants get through—particularly when there are accidents. Natural events like earthquakes, hurricanes, and tsunamis can disturb nuclear power plants and sewage systems.

The ancient Chinese believed that the Creator made two opposing forces in the creation process, called the Yin and Yang. Myths were used to explain the good force and the evil force, which could be found everywhere if one looked for it. Some countries were at peace, but others waged war, as some tribes tried to invade peaceful countries. Some countries were dry, with normal rainfall; but there were floods in others—or the reverse: no rain and drought. The Yin and the Yang can be used to explain the good and evil of the beautiful cities and towns. The good is that the cities and towns are places where food, water, shelter, jobs, and energy are all readily available. However, the evil is that water has to be diverted from rivers by building dams and reservoirs to produce electricity and provide water for people's houses and farms. Trees have to be cut down to build dwellings and to grow crops on a large scale. Animals cannot easily find food and water in their normal biomes. Roads, necessary for transportation between the cities, form a corridor for insects and animals to move along at night—with disastrous results as they are hit by vehicles travelling at high speed. Insects get confused at night when they see brightly illuminated cities.

Like the Yin and the Yang, the Creator always implements a balance to the design. In the rules of the heavens, gravity pulls matter inward in the stars and planets, while the fusion force creates the energy to push matter outward. A shining star is the result of the energy released from the actions of these two forces. The centres of billions of galaxies are believed to have gigantic Black Holes in them, where gravity is so strong that only energetic radiation, like gamma rays, can escape, and light is not energetic enough to escape. With no light escaping them, these objects are not visible even to optical telescopes. Black Holes are good because they enable the galaxy to exist. By causing the galaxy to rotate around them, time is created and stars form, giving a galaxy its unique shape. However, Black Holes have a dark side, and stars straying abnormally close to the Black Hole are torn apart and destroyed, becoming simple dust and gas while adding some mass and energy to the Black Hole so that it can continue to drive the galaxy.

In the rules of Earth, we can see the overall balance in the design. Green plants were early lifeforms that used sunlight to create food by photosynthesis. In using photosynthesis to create food, the plants removed carbon dioxide from Earth's atmosphere and added oxygen to Earth's atmosphere. If this process went unchecked, Earth's atmosphere would run out of carbon dioxide and be full of oxygen. However, the Creator introduced animals, complete with breathing apparatus to remove oxygen from the atmosphere and add carbon dioxide. This process creates a balance very similar to the concept of Yin and Yang. Also, human beings, being an extremely intelligent species with large brains, have created machines that have built cities, which use large quantities of energy. Some scientists state that fossil fuels, which have been accumulating on Earth's surface for millions of years, have nearly been completely consumed in the three hundred years since the Industrial Revolution. The expanding population results in the need for more machines, which results in more energy consumption—and more pollution, if fossil fuels are to be used. There are obviously many possibilities. People who are optimists will say there will be no Yang, and that people will build a solar- and wind-powered world, utilizing the power of the sun directly, with minimal dependence on fossil fuels. However, some people who are pessimists believe that the Yang is about to come, and that in the worst case there could be a change in the atmosphere with carbon dioxide levels rising and oxygen being depleted. Human beings will have to rely on machines, as they will be the only beings that can exist in a world with low oxygen levels. Our machines do not have to breathe to exist, they do not have to eat, and they do not need water. They will be able to survive in a polluted world. In fact, our machines will be able to exist on many other worlds in the billions of galaxies in Space. They can be designed to exist in worlds with harmful radiation, and they can easily travel in space without the need to protect sensitive organs from damage. They can withstand severe pressures and temperatures and could have the ability to live forever, if designed with suitable replaceable spare parts.

In a strange way, our machines provide a means of making the dust conscious not only on Earth or in the Milky Way Galaxy but also anywhere in the universe. Our machines provide a way of bringing our consciousness to all the galaxies. Perhaps this is the purpose of creation—an essential part of the Creator's design for the dust to become conscious in all the galaxies.

Abbreviations used

F: Fahrenheit temperature scale
C: Centigrade temperature scale
NSW: New South Wales
MIEAust: Member of the Institute of Engineers Australia
CPEng: Continued Professional Engineering Development
Ret: Retired
E: Energy m: Mass c: 300,000 kilometers per second in Einstein's equation

About the Author

Vincent J. Hyde was born in September 1954 in Calcutta, India.

While living in Calcutta, he studied at Saint Xavier's College.

He migrated to Sydney, Australia, in February 1970.

While living in Sydney, he completed his secondary education at Marcellin College and Merrylands High School, where he obtained his Higher School Certificate in 1973.

He completed an Electrical Engineering degree course at the University of New South Wales in 1979.

He completed a postgraduate Diploma in Illumination Design at the University of Sydney in 1983.

He worked as an electrical engineer at the NSW Public Works from 1979 to 2014 and retired from active engineering duties in 2014.

Since 2014, Vincent has been writing to continue his professional development.

He is a current member of the Institution of Engineers Australia MIEAust CPEng (ret).

www.ingramcontent.com/pod-product-compliance
Lightning Source LLC
Chambersburg PA
CBHW030914180526
45163CB00004B/1828